LEEDS BECKETT UNIVERSITY
Leeds Metropolitan University
17 0104940 3

Fundamentals of **Receptor, Enzyme, and Transport Kinetics**

John C. Matthews, Ph.D.
School of Pharmacy
The University of Mississippi
University, Mississippi

CRC Press
Boca Raton Ann Arbor London Tokyo

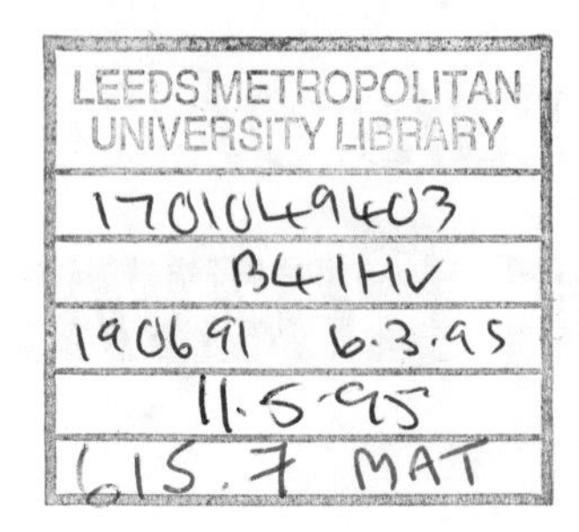

Library of Congress Cataloging-in-Publication Data

Matthews, John C. (John Charles).
Fundamentals of receptor, enzyme, and transport kinetics / by John C. Matthews.
p. cm.
Includes bibliographical references and index.
ISBN 0-8493-4426-3
1. Ligand binding (Biochemistry) 2. Enzyme kinetics. 3. Biological transport. 4. Cell receptors. 5. Pharmacokinetics. I. Title.
QP517.L54M38 1993
615'.7--dc20 92-42888
CIP

Direct all inquiries to CRC Press, Inc., 2000 Corporate Blvd., N.W., Boca Raton, Florida 33431.

International Standard Book Number 0-8493-4426-3

Library of Congress Card Number 92-42888

Printed in the United States of America 1 2 3 4 5 6 7 8 9 0

Printed on acid-free paper

PREFACE

Possibly the most valuable feature of this book is that it pulls together all of the topics of receptor, enzyme, and transport kinetics into a single, concise, easy-to-understand format, in which the terminology is consistent throughout.

This book should prove valuable as a reference source for researchers, health science and allied fields professionals who require the use of kinetic analysis of data, or for those who are evaluating kinetic data in the scientific literature. It is also intended for use as a supplementary text for graduate level pharmacology. Under limited circumstances, it can be of value for instruction at the undergraduate level.

As one might expect, this book on kinetics has many equations and graphs. The reader should not be intimidated by this fact because each concept is carefully constructed and explained, often in more than one way, with easy-to-understand analogies from daily life. Graphical representations of each key equation are presented and the important features of each graph and the related equation are explained. The organization of the material covered is from simple to complex. Complexity is built by adding new concepts, one at a time, with careful explanation, onto a firm foundation. This permits the reader to advance through the subject in a relatively painless manner.

Two useful appendices accompany the text. One is a table of equations, which outlines all of those key equations in the text and provides an indication of their uses. The second is a set of sample calculation problems and their solutions.

THE AUTHOR

John Charles Matthews, Ph.D., is an associate professor in the Department of Pharmacology in the School of Pharmacy at the University of Mississippi. He teaches biochemistry in the pharmacy degree program and biochemistry, receptor kinetics, and laboratory techniques in the pharmacology and toxicology graduate program.

Dr. Matthews' doctoral dissertation research was on the purification and characterization of a luciferase, the enzyme which catalyzes the light-emitting reaction in coelenterate bioluminescence. He was the first person to isolate and characterize calmodulin from an invertebrate. He has developed ion flux methods for the analysis of action potential sodium channels and glutamate-stimulated ion transport. His primary research interest is the elucidation of the mechanisms of age-related neurodegenerative disease.

Dr. Matthews has been the recipient of research grants from the National Institutes of Health, the National Science Foundation, The American Health Assistance Foundation, and the American Heart Association, Mississippi Chapter. He has been the author of numerous research publications and has presented his findings at numerous scientific meetings. While he has been a contributor to another volume, this is his first book.

Dr. Matthews earned his B.S. degree in chemistry and mathematics from Central Michigan University in 1969 and his M.S. and Ph.D. degrees in biochemistry from the University of Georgia in 1971 and 1976, respectively. He served as a postdoctoral research associate in biochemistry at the University of Georgia from 1976 to 1978 and as a postdoctoral research associate in pharmacology at the School of Medicine of the University of Maryland, from 1978 to 1980. In 1980 he moved to the University of Mississippi School of Pharmacy to assume an appointment as assistant professor of pharmacology. In 1985 he received the additional appointment as research assistant professor in the Research Institute of Pharmaceutical Sciences at the University of Mississippi. In 1987 he was promoted to his current rank of associate professor in both appointments. In the summer of 1989 Dr. Matthews was a visiting scientist at the National Center for Toxicological Research in Jefferson, Arkansas, and in 1990 he received a summer appointment as a fellow of the National Center for Toxicological Research Associated Universities.

Dr. Matthews is a member of the Society for Neuroscience, the American Society for Neurochemistry, the International Brain Research Organization, the American Association of Colleges of Pharmacy, the Sigma Xi research honorary society, the Southeastern Pharmacology Society, the South Central Chapter of the Society of Toxicology, and the Mississippi Gerontological Society. He is presently treasurer of his local chapter of Sigma Xi.

ACKNOWLEDGMENT

The author wishes to thank Dr. I. Wade Waters for inspiring him to write this book, Dr. Robert J. Fisher for a critical reading of the manuscript-in-preparation, and the numerous graduate students who served as test subjects during the development of much of this material.

TABLE OF CONTENTS

Part I
Receptors

Chapter 1
Introduction to Receptors

Chapter 2
Simple Receptor-Ligand Interactions

Chapter 3
Overview of Techniques for Direct Measurement of Receptor-Ligand Interactions

Chapter 4
Receptor-Ligand Interactions That Generate Proportional Physiological Effects

Chapter 5
Overview of Mechanisms for Coupling of Receptor-Agonist Interactions with Physiological Effects

Chapter 8
Antagonism

Part II
Enzymes

Chapter 9
Introduction to Enzymes

Part I: Receptors

Chapter 1

INTRODUCTION TO RECEPTORS

I. RECEPTOR-LIGAND COMPLEXES

The primary theoretical basis for modern pharmacology is the receptor concept. This holds that drugs, hormones, neurotransmitters, toxins, and other biologically active substances (collectively referred to here as ligands) exert their actions by way of interaction with receptors. The resulting receptor-ligand complexes, in turn, produce alterations in physiological processes. Not all biologically active substances exert their effects by interaction with receptors. An example of a class of drugs that does not directly involve receptors in producing biological actions is the gaseous general anesthetics. Halothane is a representative member of this class of drugs. These substances produce general anesthesia by dissolving in the lipid portions of cell membranes and altering the biophysical properties of those membranes. This type of action represents the exception rather than the rule.

Receptors are usually proteins and they are often incorporated into membranes or associated with membranes. Receptors also may be associated with structures other than membranes, such as DNA, or they can be "soluble". Receptors generally are large molecules while ligands generally are smaller molecules. Receptors can be thought of as having a pocket (or binding site) into which the ligand will fit with very precise interactions. A key fitting into a lock is a good analogy for a ligand fitting into its binding site on a receptor. Binding interactions between receptors and ligands show a high degree of stereospecificity as well as specificity for size, shape, charge, and chemical properties. Figure 1 illustrates some of these concepts. When the numbers on the shapes at the top of Figure 1 do not match those on the receptor, there are shape or chemical compatibility mismatches and the two structures won't fit.

II. AFFINITY, INTRINSIC ACTIVITY, AND EFFECT

In addition to a precise fit, binding of a ligand to its receptor is also the result of attractive forces between the receptor and the ligand. Much like the north pole of a magnet attracting the south pole of another magnet, the receptor and ligand attract one another. Once they have come together they have a tendency to stay together. In receptor terminology the receptor and ligand have affinity for one another. The ability of a ligand to bind to the receptor with high specificity and high affinity is not enough, by itself, to produce the desired action. The ligand also must be capable of stimulating the receptor when it binds with it. In other words, the ligand must have intrinsic activity. Many keys will fit into a lock but only a few keys are capable of unlocking that lock. If the ligand is capable of stimulating the receptor when it binds, then the ligand has intrinsic activity with that receptor and it is an agonist. The key that unlocks the lock represents the agonist in the lock and key analogy.

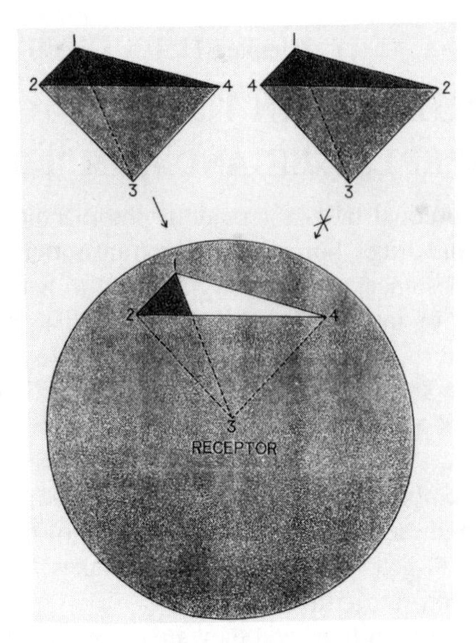

FIGURE 1

When an agonist interacts with its receptor the stimulation of the receptor initiates a sequence of events that results in a physiological response. The nicotinic acetyl choline receptor, located in the muscle cell membrane on the postjunctional side of the mammalian neuromuscular junction, is a good example for illustration of receptor coupling to a physiological response. Interaction of these receptors with acetyl choline (the endogenous neurotransmitter that stimulates them) produces changes in the receptors (probably conformational changes) that result in the opening of cation channels. The receptor-modulated cation channels are also membrane incorporated proteins that are in physical contact with and, to some extent, controlled by the receptor molecule. The channel opening event can be thought of as being mechanically triggered by the changes in the receptor protein when acetyl choline interacts with it. This would be analogous to the unlocking of the lock automatically causing the opening of the door.

When this receptor-modulated cation channel opens, sodium and calcium ions flow into the muscle cell from the outside, moving down their electrochemical gradients. This flow of positive charges into the cell reduces the inside negative electrical potential compared with the outside of the cell. This localized "depolarization" of the membrane is sensed by nearby voltage sensitive sodium channels. The voltage sensitive sodium channels respond by opening to produce a spreading depolarization of the muscle cell membrane.

In addition to the calcium that entered the muscle cell through channels from outside the cell, membrane depolarization leads to the release of calcium into the cytoplasm from the sarcoplasmic reticulum. The calcium interacts with tropomyosin and the troponin complex on the muscle contractile filaments to initiate muscle contraction. In this example tropomyosin and the troponin complex can be seen to be calcium receptors.

III. RECEPTOR STIMULATION AND PHYSIOLOGICAL EFFECT

In the example of the nicotinic acetyl choline receptor, muscle contraction is the physiological effect produced by receptor stimulation. There were several necessary intervening steps between the initial receptor-ligand interaction and the initiation of muscle contraction. Returning to the lock and key analogy, suppose that the physiological effect is the satisfaction of my hunger. I use my key to unlock the door to my house such that I can go to my kitchen, prepare a meal, and eat it. The initial receptor-ligand interaction is the insertion of the key into the lock. This is many steps removed from the final physiological response, the satisfaction of my hunger. Not all receptor-mediated physiological effects require such an elaborate series of events. Calcium binding to tropomyosin and the troponin complex of the muscle contractile apparatus leads rather directly to the physiological effect: muscle contraction.

The complexity of the processes intervening between the initial receptor-ligand interaction and the physiological response is determined, to a large extent, by the nature of the physiological response being measured. If we measure a change in membrane electrical potential near the receptor-modulated ion channel, then the receptor-ligand interaction leads rather directly to the physiological response. If we measure the ability of an animal to evade capture by a predator, an action that relies on muscle contraction (i.e., the animal's ability to run), and a complex array of other processes, then the relationship between the initial receptor-ligand event and the measured response will be much more indirect and subject to possible complications. As the intervening processes become more complex, the risk of losing the dependence of the physiological response on the initial ligand-receptor interaction increases. For example, there may be other processes that do not depend upon the particular receptor-ligand interaction which is being examined that can produce the same physiological effect. I might decide to walk down the street to a restaurant and purchase a meal instead of preparing it myself. Both alternatives will produce the same effect (satisfaction of my hunger) but the restaurant option does not require the lock and key interaction.

Returning to the nicotinic acetyl choline receptor example, we are dealing here with a transmembrane signaling system. In this system an external message, in the form of the neurotransmitter acetyl choline, results in alterations inside the cell without, itself, ever entering the cell. Signaling systems must recycle rapidly. After contracting, the muscle must relax such that it can contract again. Otherwise, the muscle will be paralyzed. In the example of the nicotinic acetyl choline receptor at the neuromuscular junction, there are several processes that function to restore the resting condition. The acetyl choline, released from the presynaptic nerve terminal, is rapidly degraded to choline and acetate by the enzyme acetyl choline esterase in the extracellular space. This reduces the concentration of acetyl choline which, in turn, causes the bound acetyl choline to dissociate from the receptors. In addition, upon interaction with acetyl choline, the receptor changes to its stimulated form in which it triggers the cation channel opening event. The receptor then automatically changes to an inactive, unstimulatable form, even though the acetyl choline is still bound. In this form the receptor is no longer capable of causing the ion channel to remain open. This process is called receptor desensitization. Once the acetyl choline has been cleared from the extracellular space by the action of acetyl choline esterase the receptor reverts to its resting form. In this form it may be stimulated upon interacting with acetyl choline.

The electrically recruited sodium channels also spontaneously inactivate; then, ion pumping enzymes in the muscle cell membrane, at the expense of metabolic energy, return the ionic balances across the various membranes to their resting conditions. This allows the contractile apparatus to relax. If there were no mechanisms available for inactivating the agonist or removing it from its receptors, nor any mechanism for receptor desensitization, the receptor-gated ion channel would remain open, the membrane would remain depolarized, the cell would exhaust its energy reserves by pumping ions out of the cell against the constant inward flow through open ion channels, and the muscle cell would remain in the contracted state.

IV. FUNDAMENTALS OF RECEPTOR-LIGAND INTERACTION

The previous discussion has served to introduce several fundamentals of receptor theory.

1. Receptor-ligand interactions involve physical contact between the receptor and the ligand.
2. When receptor and ligand come together they tend to remain together for a finite period of time because of attractive forces that operate between the pair. In other words, the receptor and ligand have affinity for one another.
3. Receptors exhibit specificity in their interactions with ligands. Only those ligands that "fit" and form an attractive force with the receptor will result in a successful interaction.
4. In order for a receptor-ligand interaction to result in the production of a physiological effect the ligand must be capable of stimulating the receptor. The ligand must be an agonist.
5. Receptor-ligand interactions are reversible.
6. Receptor-ligand interactions are concentration dependent.

V. ANTAGONISTS AND ACTIVATORS

There are many receptor-ligand interactions that do not result in receptor stimulation but are also of physiological importance. These are interactions between receptors and inhibitors (or antagonists) or interactions between receptors and stimulators (or activators). Receptor-antagonist or receptor-activator interactions are important physiologically in that they reduce or enhance (or in extreme situations, prevent or allow) receptor stimulation by the agonist. In these ways, antagonists and activators serve to regulate receptor-modulated actions. The following chapters will describe these concepts in detail, and much more concerning receptor-ligand interactions.

Chapter 2

SIMPLE RECEPTOR-LIGAND INTERACTIONS

I. LIGAND BINDING

The topic of this chapter is the interaction of a receptor with a ligand to form a receptor-ligand complex, or

$$R + L \rightleftharpoons RL \qquad \text{(Mechanism 1)}$$

Here R is the receptor, L is the ligand, and RL is the receptor-ligand complex. This is also illustrated in Figure 2.

Reaction Mechanism 1 represents a chemical reaction (or more accurately a chemical interaction) that, mathematically, is not different from any other type of chemical reaction. For example, the dissociation of an acid to form a proton and an acid anion is expressed as

$$HA \rightleftharpoons H^+ + A^- \qquad \text{(Mechanism 2)}$$

This reaction can be expressed equally well in the reverse direction

$$H^+ + A^- \rightleftharpoons HA \qquad \text{(Mechanism 3)}$$

where H^+ and A^- combine (or associate) to form HA.

II. RL LIFETIMES AND REACTION RATE CONSTANTS

When R and L combine to form RL we say that L binds to R (or R binds to L). The term 'binds' implies a reversible interaction, as does the double arrow symbol; the term also implies a semistable interaction that is more than a simple collision between the two molecules. If you throw a baseball against a wall the ball and the wall make contact with one another, but there is no lasting interaction between them. The ball bounces off the wall. If you throw a baseball to another person who is wearing a baseball glove that person will catch the ball in the glove and hold it there for a short time before throwing it back to you. In this analogy, the glove represents the receptor and the ball represents the ligand. The ball and the glove come in contact with one another in a specific fashion. The ball fits in the pouch of the glove and the glove somewhat encloses the ball. If the ball hits the glove, but the catcher fails to catch it, the collision between the ball and glove fails to result in a lasting interaction. This is analogous to a circumstance where R and L come together

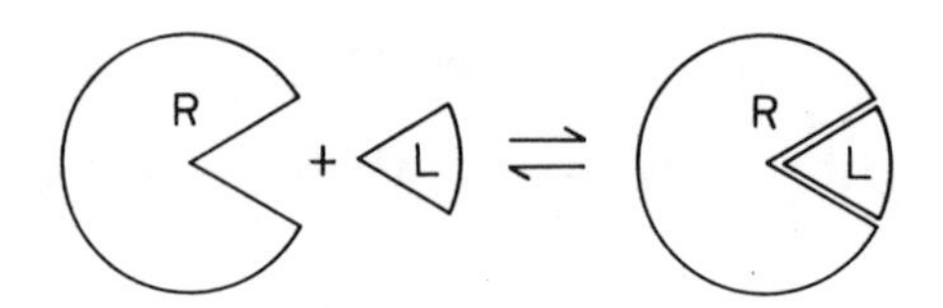

FIGURE 2

but fail to bind with one another and form an RL complex. For any single throw of the ball there can be one of at least three outcomes: (1) the catcher may fail to catch the ball, (2) the catcher may catch the ball and throw it back immediately, or (3) the catcher may catch the ball and hold it for an extended period of time before throwing it back.

The lifetime of a receptor-ligand complex, once it forms, is not a definite quantity. The complex may last a relatively short time or it may last a relatively long time. Even though we cannot assign a definite value to the lifetime of the glove-ball (receptor-ligand) complex, we can observe that it rarely lasts for less than some short but measurable time and it rarely lasts for more than some measurable, longer time. In other words, there is some predictability about the lifetime of the complex.

If we measure, a large number of times, the length of time the catcher holds the ball in the glove, we will find that for any time period we choose a constant percentage of the catches will be held in the glove for longer than this time. This constant percentage value will be independent of the number of catches made. A useful time period to choose is the time at which exactly one half of the catches are still in the glove. This represents the half-life or half-time ($t_{1/2}$) of the glove-ball (or receptor-ligand) complex. Furthermore, for any number of complexes at any time, we will find exactly one half that many remaining one half-time period later. Mathematically, we find that the lifetime of the glove-ball (or receptor-ligand) complex is defined by a simple first-order decay process. This may be expressed as

$$N = N_0 e^{-kt} \qquad (1)$$

Here N_0 is the number of complexes at time 0, N is the number of complexes remaining at time t, and k is the rate constant. When N is exactly one half of N_0, Equation 1 becomes

$$\frac{1}{2} = e^{-kt_{1/2}} \qquad (2)$$

Taking the natural log of Equation 2 gives

$$kt_{1/2} = 0.693 \qquad (3)$$

or

$$k = \frac{0.693}{t_{1/2}} \tag{4}$$

The rate constant (k) has the units of reciprocal time ($time^{-1}$). Large k values mean the reaction is fast, small k values mean the reaction is slow — k is a constant that describes the process of dissociation of the complex, therefore, it is the dissociation rate constant. A similar rate constant describes the formation (or association) of the complex and is called the association rate constant. Rate constants are only constant if the temperature is held constant. The discussion that follows in this book assumes that the temperature is held constant.

Rate constants can be added to Mechanism 1:

$$R + L \underset{k_{-1}}{\overset{k_{+1}}{\rightleftharpoons}} RL \qquad \text{(Mechanism 4)}$$

Here k_{+1} is the association rate constant for the reaction of R + L going to RL; k_{-1} is the dissociation rate constant for the reaction RL going to R + L.

III. CHEMICAL EQUILIBRIUM

It is necessary, at this point, to introduce the concept of chemical equilibrium. Equilibrium is a condition in which a chemical reaction has reached a phase where there is no net change in the concentrations of reactants and products over time. If we mix receptor and ligand together we start out with all the receptor and ligand molecules in the nonassociated form. Over time (as influenced by the rate constant k_{+1}) R and L combine to form RL and the concentration of RL increases. However, as soon as there is any RL in the mixture it begins to dissociate to regenerate R and L. It is important to note here that [R] + [RL] is a constant. If [RL] increases, then [R] must decrease by the equivalent amount. The rate of dissociation is influenced by the magnitude of k_{-1}. Progression of a reaction to equilibrium is illustrated in Figure 3. The steepness of the slope of each curve represents the rate of the reaction at any point in time. **Reaction rates are not to be confused with rate constants.** Reaction rates are the concentrations of reactants multiplied by the rate constant.

Returning to the baseball analogy, if we have 1,000 pairs of throwers and catchers on a field and we start out with only the throwers having the balls, over a short time the number of glove-ball complexes will increase from zero to some number. As the catchers catch the balls they throw them back, and so forth. If we stop all action at any instant and count the number of balls in gloves we will find that, after an initial short period, this number

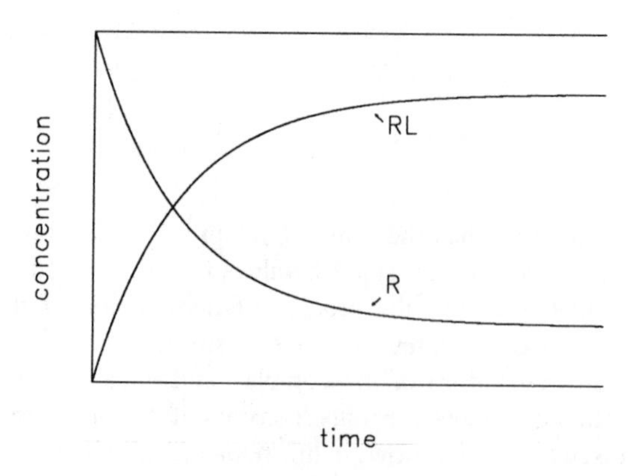

FIGURE 3

will become nearly unchanging. This is the equivalent of chemical equilibrium.

Establishment of chemical equilibrium does not mean that the reaction has stopped. Every instant, molecules of R and L are combining to form RL and molecules of RL are dissociating to form R and L. At equilibrium these reactions have reached a balance such that for each RL that dissociates an R and an L combine to replace it. The change that has occurred in the system to allow it to reach equilibrium is in the relative concentrations of reactants and products. We can see from Figure 3 that as the reaction proceeds the concentrations of the reactants decrease and the steepness of the slope of the curve decreases. The rate constant (as the term would suggest) remains constant. Thus, the reaction rate decreases in direct proportion to the concentrations of the reactants. Similarly, the rate of the reverse reaction increases with time as the concentration of the reactant for the reaction in the reverse direction increases. Eventually the reaction reaches a state where the forward and reverse reaction rates are equal. This is equilibrium.

Equilibrium translates mathematically to

$$k_{+1}\,[R]\,[L] = k_{-1}\,[RL] \tag{5}$$

The product of the reactant concentrations multiplied by the rate constant in the forward direction is equal to the product of the reactant concentrations multiplied by the rate constant in the reverse direction, when the reaction is in equilibrium. The square brackets around a term designate that term as a concentration term. Equation 5 can be rearranged to give

$$\frac{k_{+1}}{k_{-1}} = \frac{[RL]}{[R]\,[L]} \tag{6}$$

Here we see that the ratio of the rate constants for the forward and reverse reactions is equal to the product concentration divided by the reactant concentrations. In other words, the equilibrium concentrations for product and reactants are determined by the relative magnitudes of the rate constants. It is important to note here that the equilibrium concentrations of reactants and product are not necessarily equal to one another. In fact, they are generally not equal to one another. If the association and dissociation rate constants were equal to one another then, at equilibrium, [R] × [L] will be equal to [RL]. It is also important to note that, at equilibrium, less than 100% of the reactants have been converted to products. However, this number may be very near to 100%.

IV. EQUILIBRIUM AND AFFINITY CONSTANTS

We can now introduce a new term, the equilibrium constant. The equilibrium constant (K) is defined as equal to the ratio of the rate constants. We can see from Equation 6 that K is also equal to the ratio of product and reactant concentrations at equilibrium.

$$K_A = \frac{k_{+1}}{k_{-1}} = \frac{[RL]}{[R]\,[L]} \tag{7}$$

The equilibrium constant (K) is distinguished from the rate constants (k) by the convention that equilibrium constants are capitalized while rate constants are lower cased. Equation 7 describes the equilibrium constant for the reaction R + L going to RL or, it describes the association of R and L to form the RL complex. Therefore, K, in Equation 7, is an association equilibrium constant. The subscript A has been assigned to it to indicate this fact. If we were describing the reaction in the opposite direction

$$RL \underset{k_{+1}}{\overset{k_{-1}}{\rightleftharpoons}} R + L \qquad \text{(Mechanism 5)}$$

the equilibrium constant would take the form

$$K_D = \frac{k_{-1}}{k_{+1}} = \frac{[R]\,[L]}{[RL]} \tag{8}$$

Here K is now describing the dissociation of RL to form R and L. The subscript D has been assigned to indicate this fact. From Equations 7 and 8 we can see that

$$K_A = \frac{1}{K_D} \tag{9}$$

K_A and K_D describe the affinity of R and L for one another. K_A describes the tendency of R and L to come together and stay together. K_D describes the tendency of RL to dissociate.

When there are two concentration terms in the numerator and one concentration term in the denominator, as in Equation 8, the equilibrium constant will have the units of concentration (i.e., moles/liter). Similarly, when there are two concentration terms in the denominator and one concentration term in the numerator, as in Equation 7, the equilibrium constant will have the units of reciprocal concentration (i.e., liters/mole). If there are equal numbers of concentration terms in the numerator and denominator the equilibrium constant will be unitless.

From Equation 7 we see that K_A has the units of liters per mole. This is not a very useful unit to work with. One doesn't go to the laboratory and prepare solutions in liters per mole. K_D, on the other hand, has the units of moles per liter; these are much more practical units. Therefore, the equilibrium constant that we employ as the affinity constant is the K_D. When the magnitude of K_D is small the tendency of RL to dissociate to R and L is small, and RL tends to remain as a complex. When the K_D is small the affinity, of R and L for one another is large. In common usage, K_D is referred to as the affinity constant and the relationship between K_D and affinity is a reciprocal one. We will return to a discussion of the practical meaning of affinity later in this chapter.

V. RECEPTOR-LIGAND INTERACTION ENERGIES

The equilibrium constant also provides information about the energy changes involved in the receptor-ligand interaction. One way to express the second law of thermodynamics is that a chemical system will tend to go to a state of lowest free energy. In chemical reaction terms this means that a reaction will proceed in the direction that is accompanied by a loss of free energy. If RL contains less free energy than R + L, then the reaction in the direction R + L going to RL will be thermodynamically favored and at equilibrium [RL] will be greater than [R] × [L]. From Equation 8, if K_D is a small number then, at equilibrium, [RL] is large compared with [R] × [L] and the free energy content of RL is less than that of R + L. **In order for the receptor and ligand to have affinity for one another, their interaction must involve a decrease in free energy content.**

The relationship between the equilibrium constant and the free energy change is described as

$$\Delta G = \Delta G^\circ + RT\ln K_D \qquad (10)$$

Here G is the free energy, ΔG is the change in free energy for the reaction, R is the gas constant, T is the absolute (or Kelvin) temperature, and $\ln K_D$ is the natural log of the K_D. ΔG° is the free energy change associated with the reaction under standard conditions. Standard conditions are those in which all reactants and products are present at 1*M* concentration, the temperature is 298°K, and the pressure is 1 atm. Because there is no net change in the concentrations of reactants or products over time, thus no change in free energy, under equilibrium reaction conditions $\Delta G = 0$. At equilibrium, Equation 10 becomes

$$\Delta G^\circ = -RT\ln K_D \quad (11)$$

Equation 11 shows us that $\ln K_D$ is proportional to the standard free energy change for the reaction. Because the ratio of the rate constants and the equilibrium ratio of the concentrations of reactant and products (Equation 8) is equal to K_D, these quantities can also be related to the standard free energy change. If the standard free energy change for the reaction is zero then [RL] will be equal to [R] × [L] at equilibrium.

VI. THE LANGMUIR BINDING ISOTHERM

There are generally two things we want to know about a receptor: (1) the affinity of the receptor and ligand for one another, and (2) the number of receptor molecules (or more generally, the number of ligand binding sites) present. The affinity tells us how strongly the receptor and ligand interact with one another (or how stable the receptor-ligand complex is) as well as the energy change associated with that interaction. The number of receptors provides information about the magnitude of the physiological response that can be elicited. If one receptor molecule that opens one cation channel is present, we can produce a localized depolarization in one small part of the membrane. If many receptor molecules are present, scattered over the surface of the cell, we can produce a widespread depolarization of the cell membrane. As we proceed through the description of the receptor-ligand interaction the practical value of knowing these quantities will become easier to grasp. We will return to this later in this chapter.

What we need at this point is a relationship that describes the receptor-ligand interaction in terms of the K_D and the number of receptors present. Common laboratory practice is to mix the receptors and ligands in different ratios and measure the amount of RL complex at equilibrium. Generally, this is done by holding the receptor concentration constant and varying the ligand concentration. When this is done, data such as those shown in Figure 4 are obtained. From Figure 4 we see that as [L] increases (with fixed total [R]) the amount of RL increases, but not with a straight-line relationship. The curve starts out steeply at low [L] and bends, over becoming nearly horizontal at high [L]. The curve never actually becomes horizontal. It approaches horizontal as [L] approaches infinity. When [L] is infinitely large all the receptor molecules will be in the form RL. This can be seen to be the upper

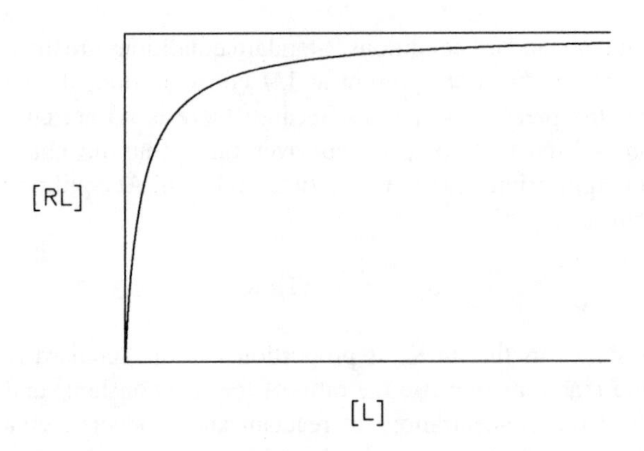

FIGURE 4

limit (the asymptote) of the curve in Figure 4. Here we must define a new term R_T,(or R total)

$$[R_T] = [R] + [RL] \tag{12}$$

The upper limit of the curve in Figure 4, where [RL] equals $[R_T]$, represents the condition where the maximum amount of ligand is bound to the receptor. We call this B_{Lmax}. Similarly the amount of ligand bound at any ligand concentration can be represented by the symbol B_L. B_L equals [RL]. Equation 12 can be rearranged to

$$[R] = [R_T] - [RL] \tag{13}$$

The equation that describes the relationship shown in Figure 4 is obtained as follows. Substituting the value of [R] from Equation 13 into Equation 8 gives

$$K_D = \frac{([R_T] - [RL])\,[L]}{[RL]} \tag{14}$$

Expanding terms yields

$$K_D = \frac{[R_T]\,[L]}{[RL]} - \frac{[RL]\,[L]}{[RL]} = \frac{[R_T]\,[L]}{[RL]} - [L] \tag{15}$$

Rearranging gives

$$K_D + [L] = \frac{[R_T]\,[L]}{[RL]} \tag{16}$$

and then

$$[RL] = \frac{[R_T]\,[L]}{K_D + [L]} \tag{17}$$

Substituting B_{Lmax} for $[R_T]$ and B_L for [RL] we have

$$B_L = \frac{B_{Lmax}\,[L]}{K_D + [L]} \tag{18}$$

Equation 18, the Langmuir binding isotherm,[1] is the equation we want. Equation 18 is identical in form to the Michaelis-Menten equation for enzyme kinetics[2] which will be described in Chapter 10. B_L is a value we measure experimentally; [L] is known since that is what we added in the laboratory. The remaining terms in the equation are the K_D, or the affinity, and the B_{Lmax}, which is the total number of receptors. Generally, in an experimental context we don't know the total number of receptors or the affinity. Equation 18 provides a means for evaluating both of these terms.

VII. THE PRACTICAL MEANING OF AFFINITY

This brings us back to the question: what is the practical meaning of affinity? K_D has the units of moles per liter, the same units as [L]. If we arbitrarily decide that one of the ligand concentrations employed to generate the data for Figure 4 was equal to the K_D, then substitution of this value into Equation 18 yields

$$B_L = \frac{B_{Lmax}\,[L]}{[L] + [L]} \tag{19}$$

This can be rearranged to

$$B_L = \frac{B_{Lmax}\,[L]}{2\,[L]} \tag{20}$$

and further to

$$B_L = \frac{B_{Lmax}}{2} \tag{21}$$

When the ligand concentration is equal to the K_D, half of the receptors will be occupied with ligand and half will be unoccupied. Thus, the affinity constant tells us the concentration of ligand that will produce one half receptor occupancy. Taking the graph from Figure 4 and finding the location of one

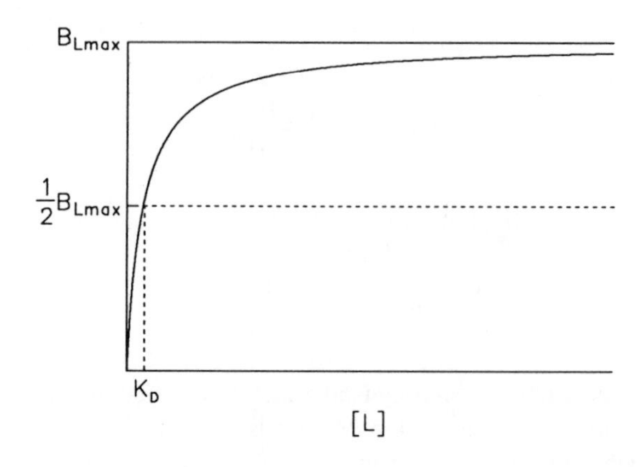

FIGURE 5

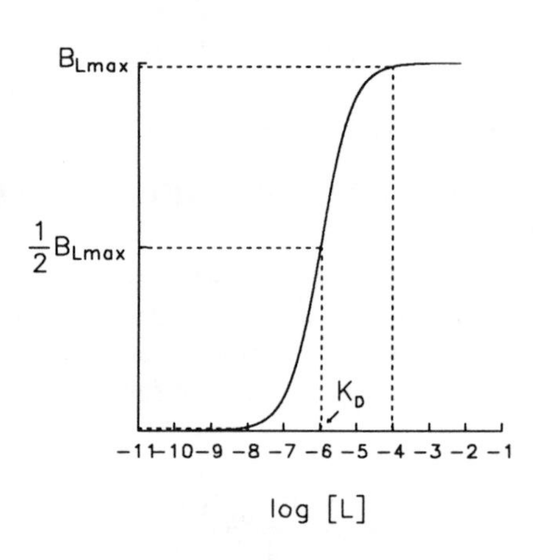

FIGURE 6

half B_{Lmax} on the ordinate, then moving horizontally to intersect the curve and down to intersect the abscissa, we obtain the value of K_D. This is shown in Figure 5.

This still doesn't fully explain the practical value of knowing the K_D. Plotting B_L vs. [L] is not a particularly useful way to plot the information. This is because everything that happens at low [L] values is squeezed together and we can't see any detail. A better way to plot the information is to plot B_L vs. log [L]. This gives Figure 6.

Plotting B_L vs. log [L] spreads out the information at low concentrations of L and compresses the information at high concentrations of L. This allows us to see in more detail the features of the relationship between B_L and [L]. One striking feature of the relationship is that only within a relatively narrow range of L concentrations is there any substantial influence of [L] on the

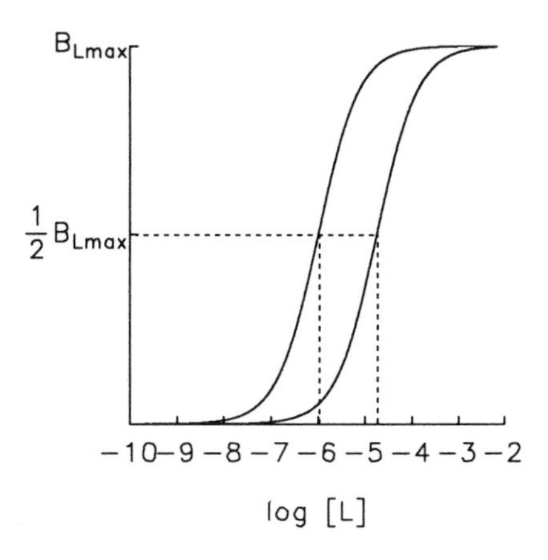

FIGURE 7

amount of RL. A second feature is that the K_D is at the exact center of this rather narrow concentration range. Within two log units (or 100-fold) of [L] above and below the K_D value we go from 99% of B_{Lmax} to 1% of B_{Lmax}. Within these four orders of magnitude of [L] (a 10,000-fold concentration range) practically all the influence of ligand concentration on receptor occupancy is accomplished. Outside of this concentration range there is very little effect of changes in the ligand concentration on receptor occupancy. Thus, the K_D value defines the effective ligand concentration range for interaction with the receptor.

It is important to note here that all binding interactions of the form described in Mechanism 1 will have exactly the shape plotted in Figure 6. The only differences from one receptor measurement to another will be the height of the curve and its left or right position on the log [L] scale. These are determined by the number of receptors present and by the K_D.

Figure 7 is a plot, on a single graph, of the association of two different ligands with different binding affinities for the same receptor. The two curves are parallel to one another, with the left-right location determined by the K_D value for the particular ligand.

If we return to the nicotinic acetyl choline receptor example, described in Chapter 1, we see that, at rest, there is little or no acetyl choline in the extracellular space near the receptors. In response to a nerve impulse acetyl choline is released into this space from the nerve terminal. It diffuses across the gap to the muscle membrane where it interacts with the receptors to initiate the muscle contraction event. Meanwhile, acetyl choline esterase degrades the acetyl choline in the extracellular space. This rapidly reduces its concentration to a level where there will be nearly no receptor occupancy.

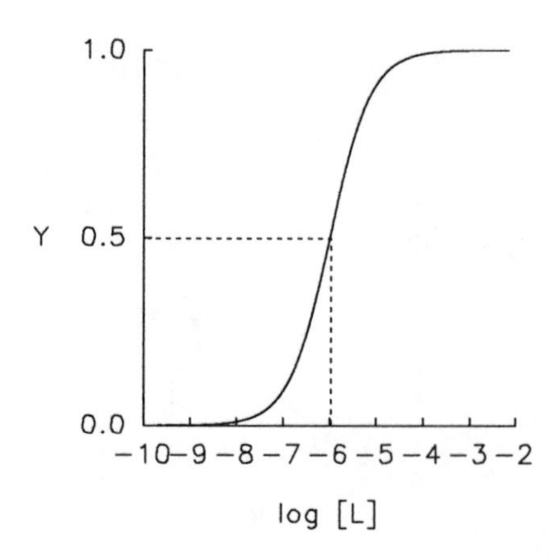

FIGURE 8

VIII. FRACTIONAL RECEPTOR OCCUPANCY

We can now define a new term, Y, the fractional receptor occupancy. Y $= B_L/B_{Lmax}$ and has values ranging from 0 to 1. Rearranging Equation 18 and substituting for B_L/B_{Lmax} yields

$$Y = \frac{[L]}{K_D + [L]} \tag{22}$$

Plotting Y vs. log [L] produces the graph shown in Figure 8. Further rearrangement of Equation 22 gives

$$Y = \frac{1}{\frac{K_D}{[L]} + 1} \tag{23}$$

Plotting Y vs. log $([L]/K_D)$ gives the graph shown in Figure 9. Conversion of the ordinate value to Y has the effect of transforming all receptor relationships to the same scale (0 to 1). In other words, it normalizes the ordinate scale. This normalization of the ordinate scale removes differences in height due to differences in the number of receptors present. The additional conversion of the abscissa to log $([L]/K_D)$ has the effect of normalizing the abscissa scale. With these changes the center point of the sigmoid curve lies at 0.5 on the ordinate and at 0 on the abscissa. Additionally, Y is 0.01 when log $([L]/K_D)$ is -2 and Y is 0.99 when log $([L]/K_D)$ is $+2$.

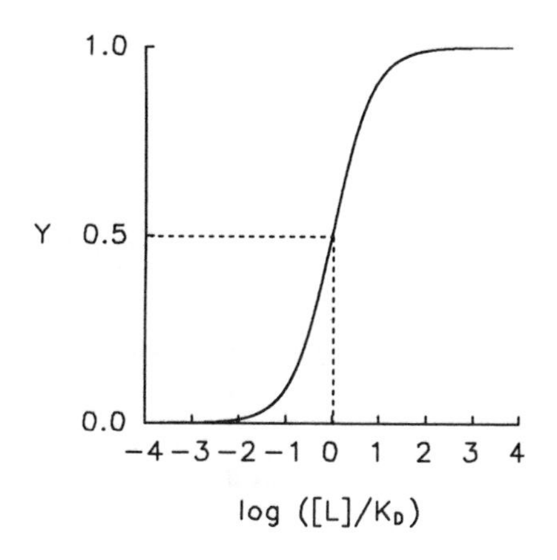

FIGURE 9

IX. LINEAR TRANSFORMATIONS

A. DOUBLE-RECIPROCAL OR LINEWEAVER-BURK EQUATION

An examination of Figures 4 and 6 reveals a potential problem with determining the values of B_{Lmax} and K_D. The problem is that with a limited amount of data, especially for very high [L], it may be difficult to obtain a reliable estimate of B_{Lmax}. If we cannot obtain a reliable measure of B_{Lmax}, then we cannot obtain a reliable measure of K_D. As is often the case, receptor ligands may be in short supply, or very expensive, or both. This may preclude obtaining sufficient numbers of data points at high [L]. Fortunately, Equation 18 can be rearranged to overcome this problem:

$$B_L = \frac{B_{Lmax}\,[L]}{K_D + [L]} \tag{18}$$

Rearranging, we have

$$\frac{B_{Lmax}}{B_L} = \frac{K_D + [L]}{[L]} \tag{24}$$

or

$$\frac{B_{Lmax}}{B_L} = \frac{K_D}{[L]} + 1 \tag{25}$$

or

$$\frac{1}{B_L} = \frac{1}{[L]}\,\frac{K_D}{B_{Lmax}} + \frac{1}{B_{Lmax}} \tag{26}$$

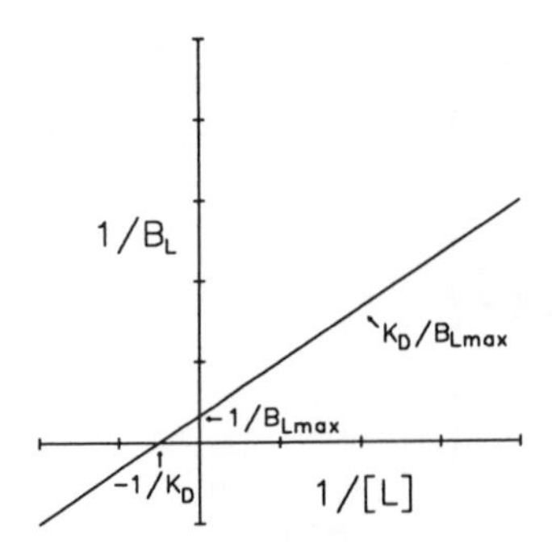

FIGURE 10

the double reciprocal or Lineweaver-Burk equation.[3] Plotting $1/B_L$ vs. $1/[L]$ gives a straight line, as shown in Figure 10. The $1/B_L$ axis intercept is equal to $1/B_{Lmax}$ and the slope is equal to K_D/B_{Lmax}. In addition, the $1/[L]$ axis intercept is equal to $-1/K_D$. Thus, we can evaluate B_{Lmax} and K_D directly from the double-reciprocal graph.

B. THE ROSENTHAL-SCATCHARD EQUATION

There are other linear transformations of Equation 18:

$$B_L = \frac{B_{Lmax}\,[L]}{K_D + [L]} \tag{18}$$

Rearranging, we obtain

$$\frac{B_L\,(K_D + [L])}{[L]} = B_{Lmax} \tag{27}$$

Expanding terms gives

$$\frac{B_L K_D}{[L]} + B_L = B_{Lmax} \tag{28}$$

Further rearrangment gives

$$\frac{B_L}{[L]} = -\frac{B_L}{K_D} + \frac{B_{Lmax}}{K_D} \tag{29}$$

the Rosenthal-Scatchard equation.[4,5] Plotting $B_L/[L]$ vs. B_L yields the graph in Figure 11. Here the slope is $-1/K_D$, the $B_L/[L]$ axis intercept is B_{Lmax}/K_D, and the B_L axis intercept is B_{Lmax}.

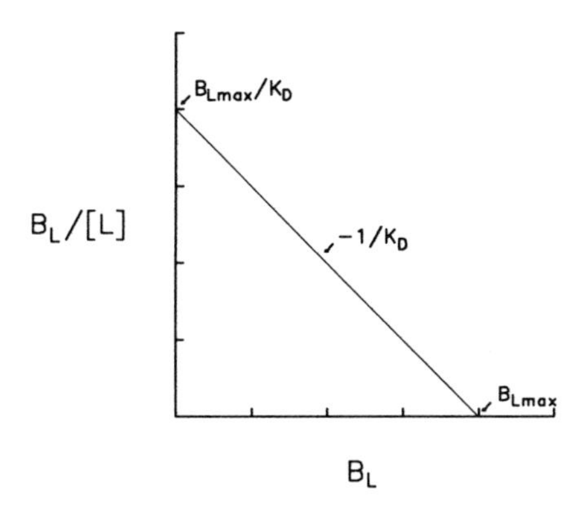

FIGURE 11

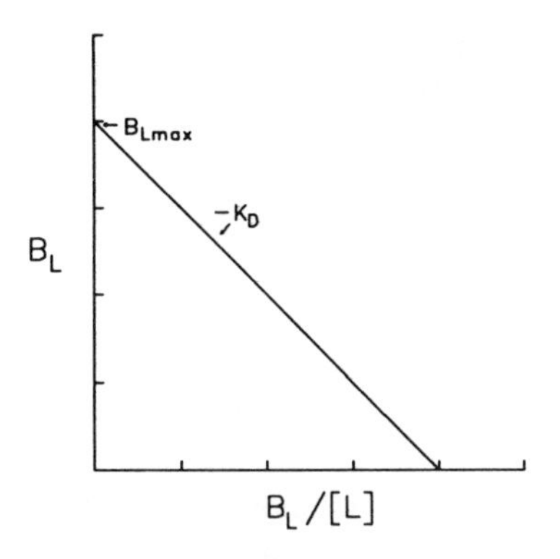

FIGURE 12

C. THE EADIE-HOFSTEE EQUATION

An alternative rearrangement of Equation 28 gives

$$B_L = -\frac{B_L}{[L]} K_D + B_{Lmax} \tag{30}$$

the Eadie-Hofstee equation.[6,7] Plotting B_L vs. $B_L/[L]$ yields the graph in Figure 12. Here the slope is equal to $-K_D$ and the B_L axis intercept is equal to B_{Lmax}. The plots from the Rosenthal-Scatchard and the Eadie-Hofstee relationships differ only from the standpoint that the axes are reversed. However, the Eadie-Hofstee plot provides for a more direct evaluation of the K_D and B_{Lmax} values.

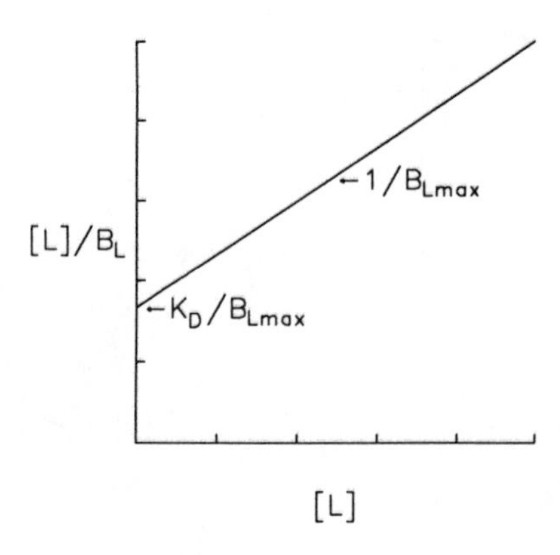

FIGURE 13

D. THE WOOLF EQUATION

Multiplying both sides of Equation 26 by [L] yields

$$\frac{[L]}{B_L} = \frac{[L]}{B_{Lmax}} + \frac{K_D}{B_{Lmax}} \tag{31}$$

the Woolf equation[8]. Plotting $[L]/B_L$ vs. [L] yields the graph in Figure 13. Here the slope is equal to $1/B_{Lmax}$ and the $[L]/B_L$ axis intercept is equal to K_D/B_{Lmax}.

The Lineweaver-Burk transformation has been the most extensively used of the four linear transformations presented above. It is useful since it is the most widely recognized and understood of the linear transformations. The Woolf equation has been the least used, but it is probably the best linear form for predicting the amount bound from a given ligand concentration. The Rosenthal-Scatchard and Eadie-Hofstee transformations are particularly useful for detecting instances where there is more than one type of binding site for the ligand at different affinities together in a preparation, or where there is cooperativity. In these situations the plots will be nonlinear. Such situations will be covered in Chapter 7.

X. WEIGHTING ERRORS

All experimental measurements are subject to some degree of error no matter how carefully the measurements are made. Each of the linear transformations of Equation 18 has the problem of weighting these errors disproportionately across the concentration range employed. This phenomenon is most easily seen with the Lineweaver-Burk transformation. Figure 14 shows that low values of [L] plotted as 1/[L] give large numbers, and errors in the

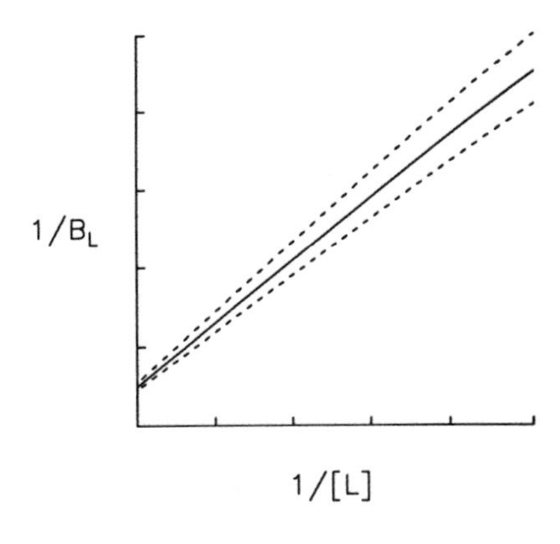

FIGURE 14

amount bound, indicated by the dashed lines, are similarly magnified. Because of this magnification, an error at low [L] can affect the slope of the line and the value obtained for K_D much more than an error of similar magnitude at high [L]. The Lineweaver-Burk transformation gives the largest weighting errors of the four linear transformations, while the Eadie-Hofstee transformation gives the smallest. Equation 18 (the Langmuir binding isotherm) does not produce inherent weighting errors. The common availability of appropriate computer programs makes nonlinear fitting to Equation 18 the most reliable means of evaluating experimental data for B_{Lmax} and K_D.

If binding data are obtained by employing radioisotope-labeled ligands, there can be inherent weighting problems, even when the data are fitted to Equation 18. These weighting problems, due to the use of radioisotopes, will be described in the next chapter.

Chapter 3

OVERVIEW OF TECHNIQUES FOR DIRECT MEASUREMENT OF RECEPTOR-LIGAND INTERACTIONS

I. RECEPTOR PREPARATIONS

Most of the commonly used techniques now available for directly measuring receptor-ligand interactions involve the use of radioisotopes. A preparation that contains receptor sites of interest is incubated with a solution that contains the radiolabeled receptor ligand. The bound ligand (RL complex) is physically separated from the remaining unbound ligand and the amount bound is then quantified by measuring the amount of radioactivity in the bound fraction. Receptors are often membrane incorporated, membrane associated, or in some other particulate form. When this is true, the physical separation of the receptor-ligand complex from the unbound ligand is much simpler than when the receptor is in ''solution''.

Cells in culture, tissue slices, tissue homogenates (and membrane fractions isolated from them), and purified receptors incorporated into artificial membranes (liposomes) are some commonly employed membranous preparations for ligand binding studies of receptors. Tissue slices anchored in an incubation vessel or cells growing anchored to a culture dish allow separation of unbound ligand by simply decanting the ligand solution. Membrane homogenates or liposomes are incubated with the ligand as particulate suspensions. Separation is accomplished by collecting the membrane particles on a filter or by centrifuging them out of suspension.

When the receptor is ''soluble'' more elaborate means of separation of bound from unbound ligand are required. Chromatographic separations and immunochemical precipitation of the receptor or unbound ligand are commonly employed methods. Alternatively, the ''soluble'' receptor can be trapped in a gel, such as those employed for electrophoresis, and handled like a tissue slice. Polyacrylamide and starch gels are among those commonly employed. Membrane homogenates, liposomes and soluble preparations are most often quantified by liquid scintillation counting or, depending on the isotope, by gamma counting. Tissue slices and gels will more likely be analyzed by autoradiography techniques.

II. CHOOSING RECEPTOR LIGANDS AND RADIOISOTOPES

Ideally, the ligand employed will bind to the receptor with both high affinity and high specificity. The radioisotope used for labeling of the ligand will, ideally, have a decay half-life of at least a few days and less than

1×10^6 years. Furthermore, when possible it is desirable that the radioisotope be attached to the remainder of the ligand by way of a stable chemical bond. The most commonly used isotopes for radiolabeling of receptor ligands are ^3H, ^{14}C, ^{32}P, ^{35}S, ^{36}Cl, and ^{125}I. ^3H, ^{14}C, ^{32}P, ^{35}S, and ^{36}Cl are beta emitters with half-lives of 12.26 years, 5730 years, 14.3 days, 88 days, and 310,000 years, respectively. ^{125}I is a gamma emitter with a half-life of 60 days. Each of these isotopes can be substituted for its nonradioactive counterpart in any molecule in which that element normally occurs without substantially altering the chemical properties of the molecule. In addition, each of these isotopes can be chemically attached to various types of molecules. Often this attachment can be accomplished in ways such that the ability of the labeled molecule to interact with the desired receptor is not seriously altered.

III. QUANTIFICATION OF RADIOACTIVITY

The three major methods of detecting and quantifying radioisotopes are autoradiography, liquid scintillation counting, and gamma counting. Of the isotopes listed above, only ^{125}I can be detected and quantified by gamma counting. Each of the listed isotopes is suitable for autoradiography and liquid scintillation counting. Gamma counting entails placing the radioactive sample in a well in a potassium iodide crystal that has been doped with a scintillator. The scintillator is a substance that absorbs the energy from the gamma radiation impinging upon it and reemits that energy as light of a much longer wavelength. The reemitted light can be detected and converted to an electronic signal by an array of photomultiplier tubes that surround the crystal. Gamma emissions are highly energetic and, therefore, gamma radiation is penetrating.

Beta emissions are of much lower energy and, for the most part, are not capable of any deep penetration. Therefore, the scintillator needs to be in close contact with the beta emitter when it disintegrates. This is accomplished by dissolving the sample in a solvent that contains the scintillator (called a scintillation cocktail) and counting in a liquid scintillation counter.

Autoradiography is normally done with tissue slices, or gels, or other preparations of similar geometries, i.e., flat and thin. These can be incubated with the radiolabeled ligand, decanted (and sometimes washed) to remove unbound ligand, and then placed in contact with a photographic emulsion. The photographic emulsion will be exposed by the radiation emitted as the isotope decays.

Each nucleus that disintegrates emits a single beta particle or a single gamma photon. Each of these types of emission has the potential to produce a single photon of light from the scintillator. Each photon of light, in turn, has the potential to produce an electronic signal, or a count. Generally, the processes of capturing the energy emitted by radioactive decay and converting it to counts are not 100% efficient. Therefore, counting efficiencies must be determined and used to convert counts per second (CPS), or counts per minute

(CPM) to disintegrations per second (DPS), or disintegrations per minute (DPM). Disintegrations per second are also known as Becquerels (Bq); 1 DPS is equal to 1 Bq. Similarly, 1 DPM is equal to 60 Bq. Another useful unit of radioactivity is the Curie (Ci); 1 Ci is equal to 1.33×10^{14} Bq.

IV. EQUILIBRIUM AND DYNAMIC BINDING MEASUREMENTS

Binding measurements may be done under equilibrium or dynamic conditions such that association and dissociation rate constants can be measured. Equilibrium binding measurements involve incubating the receptor and ligand together for a sufficient length of time such that the system can come to equilibrium. Then the amount of bound ligand is measured. Normally, equilibrium binding will provide the desired information and equilibrium binding measurements are generally more straightforward. Measurements of the association and dissociation rate constants provide alternative means of determining the affinity constant. These are sometimes employed to check the accuracies of K_D values obtained from equilibrium binding measurements.

Dynamic measurements of rate constants employ mixing of the receptor and ligand together and measuring the amount bound at various time points during the approach to equilibrium. After the system has come to equilibrium, the free ligand is rapidly removed or diluted. The amount bound at various time points is then determined as the system approaches a new equilibrium condition. Equation 1 is used to evaluate the rate constants from dynamic binding measurements:

$$N = N_0 e^{-kt} \qquad (1)$$

V. FAST AND SLOW RECEPTOR-LIGAND INTERACTIONS

Determination of the association and dissociation rate constants will also provide information about the length of time necessary for the receptor and ligand interaction to come to equilibrium. If both the association and dissociation rate constants are small numbers, it means that both reactions are slow. If both rate constants are large numbers, it means that both reactions are fast. Receptor-ligand interactions can be qualitatively classified as fast or slow, based upon the length of time necessary for the system to reach equilibrium. Ultimately this is dependent upon the relative magnitudes of the rate constants. Since the affinity constant is the ratio of the rate constants it is possible to have fast or slow receptor-ligand interactions with the same K_D value.

When working with a fast receptor-ligand interaction it is possible that a substantial amount of bound ligand can dissociate from the receptor during the step where the bound ligand is separated from the free ligand. As long as the bound ligand remains in contact with the free ligand solution this will not be a problem since the association rate also is fast. However, if the method involves a washing step, even a short washing time can result in a relatively large loss of bound ligand. This will result in an underestimation of the amount bound. Fast interacting receptor-ligand systems require shorter incubation times to reach equilibrium. Washing, if done at all with fast ligands, is generally minimal and is done as quickly as possible.

Slow interacting receptor-ligand systems require longer incubation times to reach equilibrium. These are generally analyzed with methods that incorporate extensive washing. Washing is a valuable means of reducing nonspecific binding. If a choice is available, slow interacting ligands are preferable for ligand binding studies over fast interacting ligands of similar affinities and specificities.

VI. GENERAL CONSIDERATIONS IN RECEPTOR BINDING METHODS

A. RELATIONSHIP BETWEEN RECEPTOR AND LIGAND CONCENTRATIONS

There are some general points that need to be considered when measuring receptor-ligand interactions. The first of these is that the concentration of ligand should always be substantially larger than the concentration of receptor. If this is not true the binding will not appear saturable. Saturable binding requires that the number of receptors be much smaller than the number of ligand molecules. **The ligand concentration term ([L]) used in all the equations throughout this volume is always the unbound (or free) ligand concentration.** It is generally a good idea to correct the ligand concentration for the amount bound as part of the standard binding method. This correction becomes essential when the total ligand concentration is not substantially higher than the receptor concentration.

B. NONSPECIFIC BINDING

The second consideration is the specificity of binding. The ligand chosen must be capable of interacting with the receptor of interest with high specificity. If the binding interaction is not highly specific, the ligand will bind to alternative binding sites that may be present in the preparation. In order to be able to analyze any resulting data, there must be a way to select out only the binding that was to the receptor site of interest, i.e., the specific bound. It practically never happens that 100% of the bound ligand is specific bound and therefore it is always necessary to obtain a measurement of the nonspecific bound (NSB). NSB can be subtracted from the total ligand bound to obtain the specific bound. Nonspecific binding generally, but not always, is of lower affinity than specific binding. One chooses a ligand with high affinity such that this will be true.

To measure NSB one employs a displacing ligand. The displacing ligand may simply be the nonradiolabeled version of the same molecule or it may be an entirely different molecule. The best choice of a displacer is a ligand that is a different molecule from the radiolabeled ligand. The displacer also must have specificity for the receptor of interest. This presents a better chance of eliminating (as NSB) binding to different receptor sites in preparations that have similar or higher affinity for the radiolabeled ligand compared to the receptor site of interest.

To determine NSB, one uses a set of receptor samples containing the displacing ligand at saturating concentrations (at least 1000 times its K_D for the receptor) along with the radiolabeled ligand. The displacing ligand will occupy all the available receptor binding sites, preventing any of the labeled ligand from binding. This is competitive inhibition of binding. Competitive inhibition is described in Chapter 8. Under this condition the only radiolabel that can bind is that which binds to sites other than the receptor site, i.e., the NSB. NSB is subtracted from the total amount bound to obtain the specific bound. Total amount bound is determined under the same conditions as NSB, with the exception that the displacing ligand is omitted.

NSB may be comprised of any number of components. Radiolabeled ligands are notorious for binding to the walls of assay vessels or to filters. They may bind or adsorb to membranes or proteins in a nonspecific fashion. If there are intact membrane-enclosed compartments in the preparation the radioligand may become trapped within such spaces. Most of these modes of "binding" have almost unlimited capacity for binding the ligand and, therefore, will be unsaturable and, correspondingly, nondisplaceable.

C. AFFINITY

The third consideration is the affinity of the radiolabeled ligand for the receptor. The lower the affinity the higher the concentrations of ligand required for the binding measurement. The higher the concentration of ligand employed in a receptor binding measurement the larger the proportion of NSB in the total ligand bound. Since NSB is generally unsaturable the amount bound continues to increase as the ligand concentration increases. As is often the case with receptor sites of interest, the concentration of receptors may be very low. When this is true, and when the radioligand is of low affinity, a high concentration of ligand will be required to obtain a measurable fractional receptor occupancy. Under these circumstances the amount of specific ligand bound can become a very small fraction of the total ligand bound. Situations where NSB is 90 or 99% of total binding are easily encountered. Such conditions render the ligand useless for binding measurements since the inherent measurement error will be larger than the specific bound fraction. Some decrease of NSB and increase in receptor concentrations can be accomplished by purification of the receptor. The gains available through such approaches are limited.

D. CHOICE OF ISOTOPE FOR RADIOLABELING

The fourth consideration is the choice of an isotope for radiolabeling. The longer the decay half-life of the isotope the smaller the proportion of the total label that will disintegrate within any time period. A decay half-life of 1 min means that half of the nuclei present in any sample will disintegrate in the next minute. Thus, if we have a receptor system with a K_D of 1×10^{-6} *M* we will be working in the 1×10^{-8} to 1×10^{-4} *M* concentration range with our radiolabeled ligand. If there is one atom of radioisotope per ligand molecule then we will have total radioactivity of 6×10^{16} DPM/ml at the upper limit of the binding range. This will yield 6×10^{14} DPM bound per milliliter of incubation mixture under saturating conditions. If, however, the isotope is of a longer half-life (for example, 310,000 years for ^{36}Cl) then the number of DPM bound per milliliter of incubation mixture at saturation will be only 2500. Furthermore, the number of DPM bound per milliliter of incubation mixture at the lower end of the effective ligand concentration range (1×10^{-8} *M*) will be only 25–25 DPM is near-normal background radiation. Disintegration rates in this range are virtually impossible to quantify accurately. With this constraint, when the affinity for a ^{36}Cl labeled ligand is substantially higher than 1×10^{-6} *M*, the obtainable number of DPM becomes the limiting factor. The researcher can gain one or possibly two orders of magnitude by increasing the assay volume and increasing the number of radioactive nuclei per ligand molecule, but this is the limit.

Working with isotopes of very short decay half-lives is also highly impractical. If one does not have a laboratory physically located at the site where such isotopes are produced it will take several decay half-lives before the isotope reaches the researcher. Furthermore, it is usually necessary to accomplish organic syntheses with the isotope in order to incorporate it into the desired ligand. This means more half-times will elapse. If there is sufficient isotope remaining by this point, an additional problem remains. When a nucleus disintegrates, the molecule it was a part of is no longer the same molecule. Therefore, the concentration of ligand decreases rapidly during any measurement of binding and this must be compensated for by calculation.

E. WEIGHTING ERRORS INHERENT IN THE USE OF RADIOISOTOPES

Analysis of binding data with radioisotopes can sometimes present a problem with weighting errors that cannot be overcome simply by nonlinear fitting of Equation 18 to the data (as discussed in Chapter 2). In addition to all the other sources of error in a binding method, we have an error due to the random nature of the decay process. Any single nucleus can disintegrate in the next instant or it may not disintegrate in the next 1000 years. With a small number of nuclei this can result in a significant variation from one counting period to another. This variation will be a constant proportion of the total counts in the sample and, therefore, this component of the error increases as the radioligand concentration increases. If no corrections are made such errors will influence high radioligand concentration data points more than low concentration points. One way to overcome such an error is to set up the radioligand binding measurement in such a way that the radioisotope concentration in each sample is nearly constant. This can be accomplished by employing mixtures of radiolabeled and unlabeled ligand for the different ligand concentrations, while using a different molecule as the displacer. Alternatively, some of the nonlinear fitting programs provide a means of weighting data to compensate for this type of systematic error. Generally, weighting problems caused by the random nature of radioactive decay are not large enough to be of major concern.

VII. CALORIMETRY

Since binding interactions involve a release of free energy an alternative method for measuring ligand binding can sometimes be used. This method is calorimetry. Calorimetry measures the energy (heat) released when the receptor and ligand bind. These techniques require very sensitive equipment. Their usefulness is limited to situations where the concentration of receptor sites can be made very high.

Chapter 4

RECEPTOR-LIGAND INTERACTIONS THAT GENERATE PROPORTIONAL PHYSIOLOGICAL EFFECTS

I. RELATIONSHIP BETWEEN LIGAND BINDING AND EFFECT

We now move up to the next level of complexity and begin to discuss the relationship between fractional receptor occupancy and physiological effect. It is quite possible that a ligand can bind to a receptor and have no physiological effect. Binding alone is not the whole story. Placing a key into a lock alone is generally not sufficient to produce the desired effect, i.e., unlocking the lock and opening the door. You may have several keys on your key ring that will fit into the lock, but probably only one of them is capable of unlocking the lock. Similarly, the ligand must be capable of stimulating the receptor when it binds, such that the receptor can produce its effect. If the binding of a ligand produces receptor stimulation we say that the ligand has intrinsic activity with that receptor, and that the ligand is a receptor agonist. Ligands which bind to a receptor but don't stimulate it, i.e., those that have zero intrinsic activity, will be discussed in Chapters 7 and 8.

It is reasonable to assume, at least as a first approximation, that if a ligand is capable of stimulating a receptor, then the amount of receptor stimulation will be proportional to the amount of ligand bound, or

$$S_L = gB_L \tag{32}$$

Here S_L is the amount of receptor stimulation produced by the bound ligand and g is the proportionality constant that relates S_L to B_L. Similarly, the maximal stimulation (S_{Lmax}) is reached when B_L is equal to B_{Lmax}, or

$$S_{Lmax} = gB_{Lmax} \tag{33}$$

Combining Equations 32 and 33 we get

$$\frac{S_L}{S_{Lmax}} = \frac{gB_L}{gB_{Lmax}} = \frac{B_L}{B_{Lmax}} = Y \tag{34}$$

Returning to the lock and key analogy, unlocking the lock is generally insufficient to produce the desired effect. It is also necessary to open the door. Receptor stimulation is only a prerequisite step in the production of a physiological effect by the ligand (E_L). In the simplest case, where S_L and E_L are directly proportional to one another, then

$$E_L = hS_L \tag{35}$$

Here h is the proportionality constant that relates E_L to S_L. Similarly,

$$E_{Lmax} = hS_{Lmax} \tag{36}$$

and

$$\frac{E_L}{E_{Lmax}} = \frac{hS_L}{hS_{Lmax}} = \frac{S_L}{S_{Lmax}} = \frac{B_L}{B_{Lmax}} = Y \tag{37}$$

Thus, when the conditions of S_L proportional to B_L and E_L proportional to S_L hold, we can relate the physiological effect due to a ligand to the fractional receptor occupancy. In addition, under these conditions, the ligand concentration that gives 50% of the maximum effect (also termed the ED_{50}, for effective dose 50%) is equal to the K_D.

II. THE VALUE pL_2

We can now derive another relationship, the pL_2. Common use in the literature employs the term pD_2 (for the $-\log$ of the dose that gives one half of the maximal effect for that agonist) in place of pL_2, which is used here. The divergence from the common literature usage here is in the interest of retaining consistency in the assignment of terms. L is used for any term that refers to the agonist ligand.

Substituting from Equation 37 into Equation 25:

$$\frac{B_{Lmax}}{B_L} = \frac{K_D}{[L]} + 1 \tag{25}$$

we obtain

$$\frac{E_{Lmax}}{E_L} = \frac{K_D}{[L]} + 1 \tag{38}$$

Rearranging yields

$$\frac{K_D}{[L]} = \frac{E_{Lmax}}{E_L} - 1 \tag{39}$$

Taking the log yields

$$\log K_D - \log [L] = \log \left(\frac{E_{Lmax}}{E_L} - 1\right) \tag{40}$$

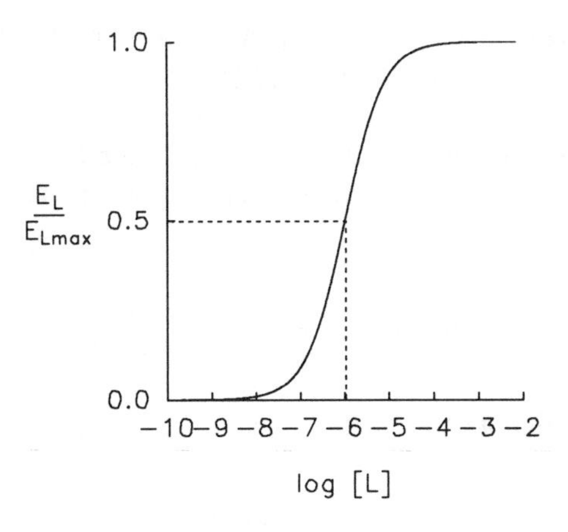

FIGURE 15

and rearranging further gives

$$-\log K_D = -\log [L] - \log \left(\frac{E_{Lmax}}{E_L} - 1 \right) \tag{41}$$

Substituting the symbol p for $-\log$ we have

$$pK_D = -\log [L] - \log \left(\frac{E_{Lmax}}{E_L} - 1 \right) \tag{42}$$

Using the same argument that we used to obtain Equation 19, when E_L equals one half E_{Lmax} then

$$-\log \left(\frac{E_{Lmax}}{E_L} - 1 \right) = -\log \left(\frac{2}{1} - 1 \right) = -\log 1 = 0 \tag{43}$$

and thus

$$pK_D = -\log [L] \text{ or } pk_D = pL_2 \tag{44}$$

pL_2 is the $-\log$ of the ligand concentration that yields one half of the maximal physiological effect for that ligand. To reiterate, when B_L is proportional to S_L and S_L is proportional to E_L, then pL_2 is equal to pK_D; otherwise they are not equal. Often we do not have any information about these proportionalities. This is one reason for using pL_2 instead of pK_D.

Figure 15 illustrates the relationship between the fractional effect and log [L].

III. pH EQUATIONS AND RECEPTOR-LIGAND INTERACTIONS

It is of interest to note here that Equation 42 can be rearranged to give an equation in the same form as the familiar Henderson[9]-Hasselbalch[10] equation for pH calculations,

$$pH = pK_a + \log\left(\frac{[A^-]}{[HA]}\right) \tag{45}$$

Rearranging Equation 42 we obtain

$$-\log [L] = pK_D + \log\left(\frac{E_{Lmax}}{E_L} - 1\right) \tag{46}$$

or

$$pL = pK_D + \log\left(\frac{E_{Lmax}}{E_L} - 1\right) \tag{47}$$

Then, since

$$\frac{E_{Lmax}}{E_L} - 1 = \frac{[R_T]}{[RL]} - 1 \tag{48}$$

and substituting for $[R_T]$ from Equation 12 we have

$$\frac{E_{Lmax}}{E_L} - 1 = \frac{[R] + [RL]}{[RL]} - 1 = \frac{[R]}{[RL]} + 1 - 1 = \frac{[R]}{[RL]} \tag{49}$$

Substituting from Equation 49 into Equation 47 yields the equivalent of the Henderson-Hasselbalch equation:

$$pL = pK_D + \log\left(\frac{[R]}{[RL]}\right) \tag{50}$$

IV. RELATIVE INTRINSIC ACTIVITY

It is quite possible that not all ligands which bind to a particular receptor have the same intrinsic activity. In other words, the same level of receptor occupancy for two different ligands that both bind to the same receptor may give different levels of stimulation. Returning to the lock and key analogy, the simplest case is one in which a key fits into a lock but will not unlock it. This would be an example of zero intrinsic activity. Receptor ligands that have zero intrinsic activity are antagonists. These will be discussed in Chapter 8. It is possible for ligands to have intrinsic activities anywhere in the range from zero to maximum. Any ligand that has greater than zero intrinsic activity with a receptor is an agonist for that receptor. Those agonists that have less than maximal intrinsic activity are termed partial agonists. As we shall see in Chapter 8, partial agonists can act as antagonists in the presence of full agonists. We all have experienced situations where our key fits into a lock and is capable of unlocking it, but we have to struggle with the key to get it to work. This would be an example of intrinsic activity greater than zero but less than maximum.

How do we get a measure of intrinsic activity? The total stimulus produced by a ligand interacting with its receptor is

$$S_L = a_L \, [RL] \, v \tag{51}$$

and similarly,

$$S_{Lmax} = a_L \, [R_T] \, v \tag{52}$$

Here a is the intrinsic activity and v is the volume of the bio-phase where R and L interact. In a situation where we have injected a ligand into an animal we don't know [RL] or $[R_T]$ and we don't know v. One might ask: isn't v simply the volume of the animal? The answer is, not necessarily, and in fact, rarely. As an example, the ligand and receptor may only interact with one another to produce a physiological effect in the extracellular spaces of the brain, and perhaps in only a small part of the brain.

The distribution of the ligand in the brain is determined by a variety of pharmacokinetic factors. These include the permeability of the ligand to the blood-brain barrier, its concentration in the circulation, and metabolic alterations of the ligand. Often, we cannot measure or even identify all the variables involved. Also, we cannot measure S_L. What we measure is E_L. If, however, we have two ligands available, which both interact with the same receptor but with different intrinsic activities, then we have a means of obtaining a relative measurement of the intrinsic activity.

$$S_{L_1max} = a_{L_1} \, [R_T] \, v \text{ and } S_{L_2max} = a_{L_2} \, [R_T] \, v \tag{53}$$

L_1 and L_2 represent ligands 1 and 2, while a_{L1} and a_{L2} are the intrinsic activities for ligands 1 and 2. Combining Equations 53 gives

$$\frac{S_{L_1max}}{S_{L_2max}} = \frac{a_{L_1} \, [R_T] \, v}{a_{L_2} \, [R_T] \, v} = \frac{a_{L_1}}{a_{L_2}} \tag{54}$$

or, the intrinsic activity ratio equals the maximum stimulus ratio.

If we have available all known ligands that interact with the receptor we can determine which of them produces the largest physiological effect. We can then assign the value of a_{max} as the intrinsic activity of the ligand or ligands in the group that produces the largest physiological effect. The relative intrinsic activity (α) can then be defined as

$$\alpha = \frac{a_L}{a_{max}} = \frac{S_{Lmax}}{S_{max}} \qquad (55)$$

Here S_{max} is the maximum known stimulation for the receptor system in question and α will have values ranging from 0 to 1 — α is sometimes referred to as the efficacy of the ligand for producing a physiological effect. This is not entirely correct. More precisely, the efficacy is proportional to α.

Again, since we are measuring the effect rather than receptor stimulation, and if effect is proportional to stimulation, then

$$\alpha^E = \frac{E_{Lmax}}{E_{max}} \qquad (56)$$

α^E is the relative intrinsic activity for producing the physiological effect and E_{max} is the maximum known effect for the receptor system in question. Since

$$\frac{S_{Lmax}}{S_{max}} = \frac{E_{Lmax}}{E_{max}} \qquad (57)$$

then

$$\alpha = \alpha^E \qquad (58)$$

and

$$\alpha = \frac{E_{Lmax}}{E_{max}} \text{ or } E_{Lmax} = \alpha\, E_{max} \qquad (59)$$

Substituting from Equation 59 into Equation 38 we obtain

$$\frac{\alpha\, E_{max}}{E_L} = \frac{K_D}{[L]} + 1 \qquad (60)$$

Rearranging gives

$$\frac{E_L}{E_{max}} = \frac{\alpha}{\frac{K_D}{[L]} + 1} \qquad (61)$$

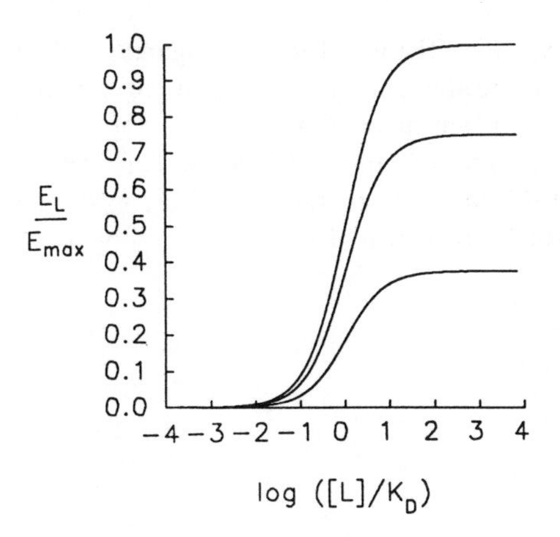

FIGURE 16

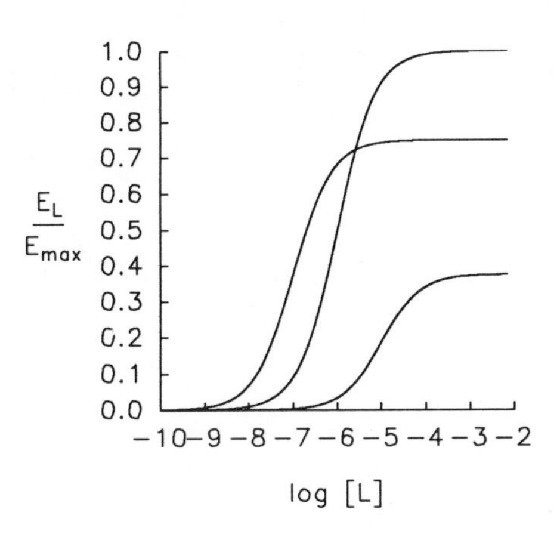

FIGURE 17

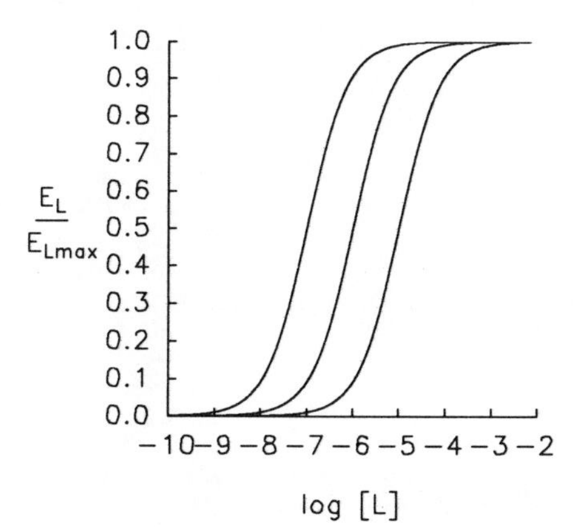

FIGURE 18

Plotting E_L/E_{max} vs. log ([L]/K_D) for three ligands with different α values gives Figure 16, where the maximum for each curve is α_L. Plotting E_L/E_{max} vs. log [L] (Figure 17) instead of E_L/E_{max} vs. log ([L]/K_D) gives a more realistic representation of the family of ligands. This is because the K_D values among a group of ligands will generally be different. Remember that plotting E_L/E_{Lmax} instead of E_L/E_{max} would produce a family of curves of the same height (Figure 18).

Chapter 5

OVERVIEW OF MECHANISMS FOR COUPLING OF RECEPTOR-AGONIST INTERACTIONS WITH PHYSIOLOGICAL EFFECTS

I. INTRODUCTION

There are many mechanisms by which receptor-ligand interactions are coupled to effector systems for the production of physiological effects. It is not the intention here to cover all such mechanisms or to go into any great detail with any of those discussed. The intent is to provide a brief overview of the types of coupling mechanisms known to occur.

A large proportion of receptors are incorporated into, or associated with membranes and, for the most part, these function in transmembrane signaling. Generally, with membrane-associated receptor systems a receptor-agonist interaction on one side of the membrane results in the production of a biochemical or biophysical change on the other side of the membrane. Some types of signal transduction systems modulated by receptors include ion channels, cyclic AMP- or cyclic GMP-dependent protein kinases, and inositol phosphate-mediated calcium release from the endoplasmic reticulum.

II. ION CHANNEL-LINKED SIGNAL TRANSDUCTION

All living cells maintain an unequal distribution of ions across their plasma membranes. Na^{+} and Ca^{++} are high in concentration outside the cell while K^{+} and Mg^{++} are high in concentration inside the cell. These ionic gradients are maintained at the expense of metabolic energy by ion pumping enzymes in the membranes. Cells use these ionic gradients for various signaling purposes. They do so by employing receptor-modulated ion channels to control (or gate) the flow of ions down their electrochemical gradients from one side of the membrane to the other. Two results of ions flowing across membranes are useful for signaling purposes. One is that changes in transmembrane ion flow result in alterations in transmembrane voltages (or polarities). There are many membrane-incorporated effector systems, particularly voltage-gated ion channels, which can sense changes in transmembrane voltage and initiate their action in response.

The second effect of transmembrane ion flow is that the ion itself, most often calcium, can act as the messenger. Calcium functions as a messenger by interacting with receptors inside the cell to initiate physiological responses. A common, but certainly not the only, cellular calcium receptor is the protein calmodulin. Calcium-calmodulin complexes interact with a variety of enzymes and other proteins in the cell to activate or inactivate them. Calcium also interacts directly with several enzymes and proteins in the cell to alter their activities.

In Chapter 1, the nicotinic acetyl choline receptor of the mammalian neuromuscular junction and its coupling to muscle contraction was described. This system is a good illustration of ion channel-mediated transmembrane signaling. Ion channels may be directly coupled to receptors, which is probably true for the nicotinic acetyl choline receptor, or they may be indirectly coupled to receptors through intervening effector systems. For example, some ion channels are modulated by G proteins similar to those to be discussed below.

III. CYCLIC NUCLEOTIDE-LINKED SIGNAL TRANSDUCTION

A. STIMULATORY RECEPTORS

Many membrane-associated receptor systems initiate their physiological effects through cyclic nucleotide-stimulated protein kinases or through direct stimulation of protein kinases without the requirement for the cyclic nucleotide intermediates. The best understood of these are the adenylate cyclase systems. Adenylate cyclase is an enzyme associated with the inner surface of the plasma membrane of many cells. Adenylate cyclase converts ATP to cyclic AMP. Cyclic AMP interacts with its receptor, a protein kinase enzyme, to activate it. The protein kinase then catalyzes the transfer of a phosphate group from ATP to a hydroxyl group of another enzyme or protein, to activate or inactivate that protein. The protein, which becomes phosphorylated by the action of the protein kinase, is invariably an important component of the physiological process being regulated. Its phosphorylation can either turn on or turn off (or, more likely, speed up or slow down) that entire physiological process.

The activity of the adenylate cyclase enzyme is regulated by a receptor on the outside of the cell membrane through an intervening G protein. When the agonist binds to its receptor the stimulated receptor-agonist complex causes the G protein to exchange its bound GDP molecule for a molecule of GTP from inside the cell. The G protein-GTP complex, in turn, interacts with the adenylate cyclase to activate it. The G protein, which is a GTPase enzyme, then hydrolyzes the GTP to GDP. This terminates the ability of the G protein to activate adenylate cyclase. Meanwhile, cyclic nucleotide phosphodiesterase enzymes and protein phosphatase enzymes in the cell function to reverse the effects of the underlying receptor stimulation. They do so by converting cyclic AMP to AMP and by removing the phosphate groups from the phosphorylated proteins. As was true with the nicotinic acetyl choline receptor, the extracellular agonist will only be available to interact with its receptor for a short period of time. In addition to enzymatic degradation of the agonist, several cellular uptake processes are important for rapidly clearing receptor agonists from the extracellular space as a means of terminating receptor stimulation.

B. INHIBITORY RECEPTORS

The adenylate cyclase receptor-linked systems have an additional feature, that of also being coupled to an inhibitory receptor. The inhibitory receptor interacts with its agonist to cause an inhibitory G protein to exchange its bound GDP molecule for GTP. The inhibitory G protein-GTP complex then produces inhibition of the adenylate cyclase enzyme, instead of activation. This is functional antagonism and will be discussed again briefly in Chapter 8.

IV. PHOSPHATIDYL INOSITOL-LINKED SIGNAL TRANSDUCTION

Some membrane-associated receptor systems accomplish transmembrane signaling through the production of inositol triphosphate (IP_3) and diacylglycerol by way of G protein-mediated stimulation of the enzyme phospholipase C. Phospholipase C, an enzyme associated with the inside surface of the cell membrane, catalyzes the hydrolysis of a minor component of the phospholipid membrane, phosphatidylinositol-4,5-diphosphate. The products of this hydrolysis are one molecule of IP_3 and one molecule of diacylglycerol. The water soluble IP_3 diffuses through the cytoplasm to the endoplasmic reticulum membrane. Here it interacts with its receptors to stimulate the release of calcium into the cytoplasm from stores in the endoplasmic reticulum. The calcium, in turn, interacts with cytoplasmic calcium receptors, such as calmodulin or troponin, to produce the physiological effect. The lipid soluble diacylglycerol remains in the plasma membrane where it interacts with and activates a family of protein kinase enzymes. These protein kinases, in turn, catalyze the phosphorylation of a variety of other enzymes and proteins to alter their activities, and thus contribute toward producing the physiological effect. IP_3 and diacylglycerol are rapidly recycled in the cell to reform phosphatidylinositol-4,5-diphosphate. Other inositol phosphate metabolites, produced as intermediates in the recycling reactions, also may be important as intracellular signals.

V. STEROID RECEPTORS

There are many receptors that function by means other than transmembrane signaling. We have already talked about IP_3 receptors, protein kinases, calmodulin, and the troponins, which fall into this category. Steroid receptors represent an additional example. Many steroid receptors exist whose function is to modulate the expression of information stored in DNA molecules. Steroids are hormones that are often capable of diffusing across membranes unassisted, and thus are capable of interacting with receptors inside cells. Generally, steroid receptors are "soluble" components of the cell cytoplasm. These form a complex with their steroid agonists and then diffuse into the cell nucleus. Here they bind to DNA in such a fashion that they can regulate the expression of predetermined gene products.

Chapter 6

OVERVIEW OF TECHNIQUES FOR MEASURING RECEPTOR-AGONIST INTERACTIONS THROUGH THEIR PHYSIOLOGICAL EFFECTS

I. INTRODUCTION

Ligand binding measurements, as discussed in Chapter 3, are incapable of telling us anything about the physiological effects of a receptor-ligand interaction. As we shall see in Chapter 8, there are many receptor-ligand interactions (antagonism) that do not result in receptor stimulation. Technically, these cannot produce a physiological effect. We say technically because they do produce physiological effects through antagonism of agonist actions. Therefore, it is necessary to be able to measure the physiological effect of a receptor-ligand interaction in order to be able to correlate that interaction with the physiological effect.

There are many ways to measure physiological effects of receptor-ligand interactions. Methods which may be employed for the measurement of physiological effects due to receptor-agonist interactions range from conscious live animals, through organ and tissue preparations at various levels of isolation, through cells in cultures and isolated membrane preparations, to isolated receptor-effector systems at the molecular level.

II. EFFECTS ON WHOLE ANIMALS

With a whole animal, the agonist is administered by any of several routes that can include various routes of injection, feeding, inhalation, or absorption through epithelial tissues. The animal's response to the agonist is then measured. The responses measured may be relatively straightforward and noninvasive, such as changes in core body temperature, blood pressure, or heart rate, or they may be more complex such as changes in the behavior or metabolic status of the animal. Simply stated, any change resulting from administration of a receptor agonist can be used as a measure of effect.

Problems encountered in working with whole animals include difficulties in determining the concentration of agonist at the receptor and the continuous change in the concentration of agonist at the receptor. Once the agonist has been introduced into the animal it must be distributed to the location of the receptors. This distribution often involves the crossing of physical barriers such as membranes, traveling in the blood or lymphatic system, binding to tissue or blood components other than the receptors that produce the effect being measured, and unequal uptake by various tissues in the body. Many, if not most, substances that function as receptor agonists are also subject to metabolic alterations once they enter the animal. Generally, such alterations reduce or eliminate the receptor agonist activity. Sometimes, however, metabolic alterations are required to convert the administered compound to an active agonist or they may increase its activity. In addition, the agonist is also subject to elimination from the animal by various routes of excretion.

These and other processes serve to alter the concentration of the agonist at the receptor over time. This makes it difficult, if not impossible, to determine the exact concentration of agonist at the receptor. Therefore, with whole animal experiments the dose administered (usually in milligrams of agonist per kilogram of animal) is employed for calculations in place of the concentration of agonist at the receptor. This is also an important reason for employing the pL_2 in place of the pK_D. Based on the considerations outlined above, the administered dose may not be similar to the concentrations required to produce binding interactions with the receptor.

The route of administration and the time after administration when the measurement is performed are also important parameters for comparing results. Another complication in the measurements of agonist effects on whole-animal preparations is that the effect measured may not occur until long after the agonist has been eliminated from the animal by metabolism and excretion. The lock and key analogy is appropriate here for illustration. I use my key to unlock the door to my house so that I can get inside. Once inside I go to my kitchen, prepare myself a meal and eat to satisfy my hunger. The unlocking of the lock, which was the event that initiated the entire process of producing the physiological effect of satisfying my hunger, occurred early in the time sequence. By the time I have completed my meal, the key has been back in my pocket and the door has been relocked for a relatively long time.

III. ISOLATED ORGANS AND TISSUES

Isolated organ and tissue preparations have been extensively used for measuring effects due to receptor-agonist interactions. These methods usually involve removing the organ or tissue from the donor animal and mounting it in a device that bathes it in an artificial medium and allows measurement of the effects. Isolated heart preparations are a good example. The spontaneously beating heart is mounted in an oxygenated bath in a bathing solution that provides various salts, buffers, and sometimes glucose. The isolated heart is connected to a force transducer that allows the measurement of the strength and rate of contraction, and possibly other parameters. Increasing concentrations of the agonist are then placed in the bath and the effects are determined.

Isolated organ and tissue preparations eliminate excretion and greatly reduce problems associated with the distribution and metabolism of the agonist compared with those encountered in the whole animal. This allows a more accurate estimation of agonist concentrations at the receptor. Isolated organ and tissue preparations also eliminate many complications such as feedback effects and hormonal and neural controls that originate outside of the tissue or organ.

IV. TISSUE SLICES, CELL CULTURES, AND MEMBRANE VESICLES

Tissue slices, cell cultures, and membrane preparations further reduce potential problems with distribution and metabolic alteration of the agonist. These preparations allow estimation of agonist concentrations at the receptor as accurately as can be made with ligand binding methods. Furthermore, it is often possible to make ligand binding and effect measurements on identical preparations such that the results will be directly comparable. As a general rule, as the preparation becomes less complex the nature of the effect that is being measured also must become less complex. For example, one might measure the increase in concentration of cyclic AMP in a cell culture preparation as compared with a change in behavior in a whole-animal preparation.

V. RECEPTOR-EFFECTOR SYSTEMS AT THE MOLECULAR LEVEL

Isolated receptor-effector systems can sometimes be analyzed at the molecular level. For example, calmodulin is an intracellular calcium receptor that interacts with several enzymes and other macromolecules in the cell to alter their activities. If one measures the activity of an isolated enzyme, which is regulated by calcium-calmodulin, one has a receptor-effector system fully at the molecular level. Enzymes can be stimulated directly by ligands, as discussed further in Chapter 11. These represent the simplest receptor-effector systems.

Chapter 7

RECEPTOR-LIGAND INTERACTIONS THAT ARE DISAPROPORTIONATE WITH THEIR PHYSIOLOGICAL EFFECTS

I. INTRODUCTION

The next level of complexity to be considered is that of receptor-ligand interactions that are disproportionate with the physiological effect they produce. So far, all the situations we have dealt with are those in which S_L is proportional to B_L and E_L is proportional to S_L. There are situations where S_L is not proportional to B_L or E_L is not proportional to S_L, or both.

II. SPARE RECEPTORS

An often encountered example of such disproportionality is that of spare receptors. Consider a cell membrane that has receptor-gated calcium channels in it. The purpose of these channels is to permit the controlled entry of calcium into the cell cytoplasm from outside the cell. Once inside the cell, the calcium interacts with a variety of calcium receptors to initiate its physiological effect. In order for the calcium to interact with its receptors inside the cell its concentration must be elevated sufficiently to put it in the range where it can have a significant effect on fractional receptor occupancy. Let us suppose that opening half of the calcium channels will permit the entry of enough calcium in a short enough time to saturate all the calcium receptor sites inside the cell. This will produce the maximum obtainable physiological effect. If more of the calcium channels are opened by the addition of more ligand no further effect will be produced. All the receptor-gated calcium channels over 50%, in this example, are spares.

Returning to the lock and key analogy, suppose I go home to prepare myself a meal to produce the physiological effect of satisfying my hunger. When I arrive home I insert my key into the door lock, unlock the lock, open the door, enter the house, go to the kitchen, prepare the meal, and eat. All the steps leading to satisfying my hunger were initiated by the key fitting into the lock and being able to unlock it. Now suppose that my house has three doors, all of which can be unlocked with the same key. In order to achieve the maximum effect I need only unlock and open one of these doors. It doesn't matter which one. They all allow me to gain access to the kitchen. In this example, the two other doors are spares. When it is only necessary to stimulate a few of many receptors to produce the maximum achievable physiological effect E_L is not proportional to S_L.

In a situation where we have spare receptors, full agonists and partial agonists are likely to produce the same maximal effect. This is shown in

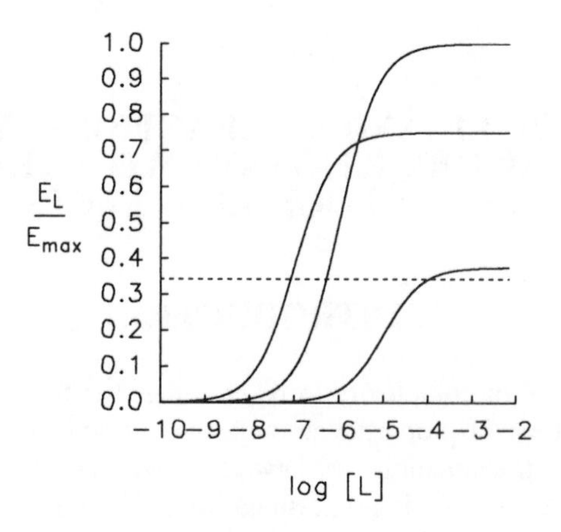

FIGURE 19

Figure 19. The dashed line represents the E_{max} when there are spare receptors. It must be noted here that the original graph cannot be generated as it appears when spare receptors are present. It will be necessary to stimulate a larger proportion of the total number of receptors to produce the maximum effect with a partial agonist than with a full agonist. However, if there are enough receptors present the maximum effect can be achieved. A further examination of Figure 19 reveals that the presence of spare receptors reduces the agonist concentration range required to produce the effect (spare receptors steepen the log [L] vs. effect curve).

III. HOMOTROPIC COOPERATIVITY AND THE HILL EQUATION

The second circumstance that can lead to disproportionality between receptor occupancy and physiological effect is when single receptors have multiple binding sites for the same ligand. These binding sites may have fixed affinities, or the binding of the ligand to one site may influence the affinity for binding of the ligand to the remaining sites. This is homotropic cooperativity. It is impossible to distinguish the two different mechanisms from kinetic data alone. When homotropic cooperativity is operative S_L is related to E_L, but not by a simple proportionality. It may be necessary to occupy and stimulate several of the ligand binding sites before a physiological effect can be produced.

The Hill relationship[11] is useful to analyze for homotropic cooperativity. Beginning with the relationship derived in Chapter 2 for fractional receptor occupancy

$$Y = \frac{1}{\frac{K_D}{[L]} + 1} \tag{23}$$

which describes Mechanism 1

$$R + L \rightleftharpoons RL \qquad \text{(Mechanism 1)}$$

we need an expression that represents the more general circumstance of more than one ligand binding site per receptor molecule:

$$R + nL \rightleftharpoons R(L)_n \qquad \text{(Mechanism 6)}$$

Here n is the number of binding sites on a single receptor molecule. Mechanism 6 also can be expressed as

$$R + L \rightleftharpoons RL + L \rightleftharpoons RLL + L \rightleftharpoons RLLL \text{ etc.} \qquad \text{(Mechanism 7)}$$

For this mechanism

$$K_D = \frac{[R]\,[L]^n}{[R(L)_n]} \tag{62}$$

The K_D in Equation 62 is actually the sum of all the equilibrium constants for each of the reactions in Mechanism 7. Equation 62 can be rearranged to give

$$[R] = \frac{K_D[R(L)_n]}{[L]^n} \tag{63}$$

Here

$$[R_T] = [R(L)_n] + [R] \tag{64}$$

and

$$Y = \frac{[R(L)_n]}{[R_T]} = \frac{[R(L)_n]}{[R(L)_n] + [R]} = \frac{[R(L)_n]}{[R(L)_n] + \dfrac{K_D\,[R(L)_n]}{[L]^n}} \tag{65}$$

or

$$Y = \frac{1}{\dfrac{K_D}{[L]^n} + 1} \tag{66}$$

Equation 66 compares to Equation 23. Plotting Y vs. log [L] from Equation 66 for n values of 1 and 2 yields Figure 20. Figure 20 shows that the presence

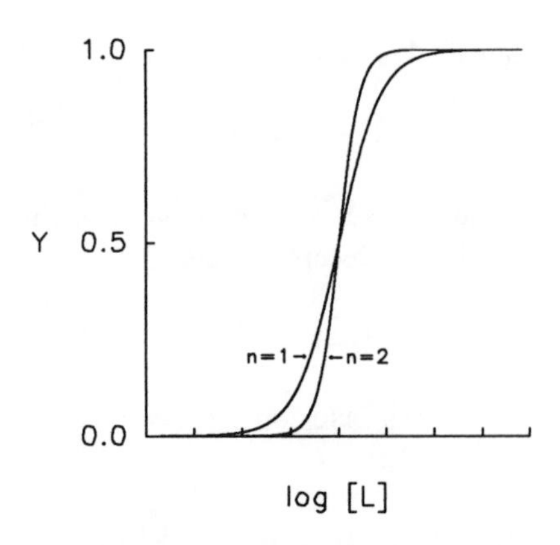

FIGURE 20

of multiple binding sites decreases the effective ligand concentration range necessary for the production of changes in fractional receptor occupancy. The sigmoid curve becomes steeper as n increases.

We want to use Equation 66 to obtain an expression for the ratio of occupied receptor sites to those that are unoccupied ($[R(L)_n]/[R]$). In terms of fractional receptor occupancy we want an expression for Y/(1-Y). When we invert Y/(1-Y) we have

$$\frac{1-Y}{Y} = \frac{1}{Y} - \frac{Y}{Y} = \frac{1}{Y} - 1 \tag{67}$$

Then, rearranging Equation 66 gives

$$\frac{1}{Y} = \frac{K_D}{[L]^n} + 1 \text{ or } \frac{1}{Y} - 1 = \frac{K_D}{[L]^n} \tag{68}$$

Inverting again gives

$$\frac{Y}{1-Y} = \frac{[L]^n}{K_D} \tag{69}$$

Then, as is often the case, we want an equation that describes a straight line. Taking the log of Equation 69 yields

$$\log\left(\frac{Y}{1-Y}\right) = n \log [L] - \log K_D \tag{70}$$

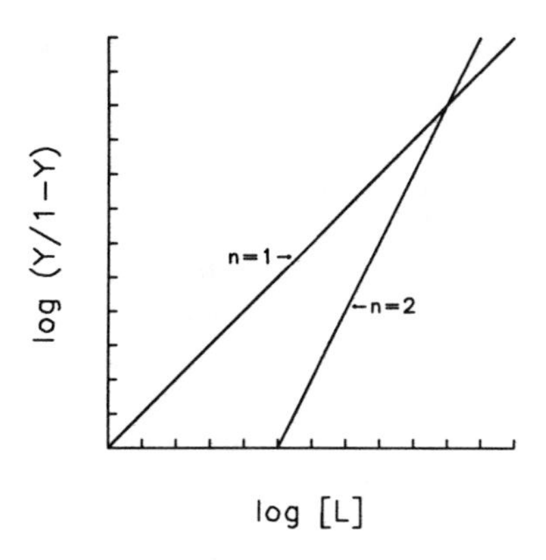

FIGURE 21

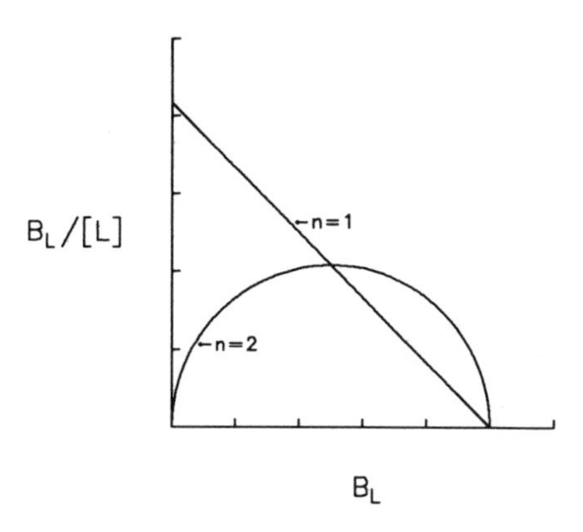

FIGURE 22

the Hill equation. Plotting log (Y/1-Y) vs. log [L] gives straight lines with slope n. This is shown in Figure 21. If n is greater than 1 there is more than one binding site for the ligand on the receptor.

Scatchard plots are also useful for detecting and analyzing for the presence of homotropic cooperativity. Equation 65 can be rearranged to its Scatchard equivalent form

$$\frac{B_L}{[L]^n} = -\frac{B_L}{K_D} + \frac{B_{Lmax}}{K_D} \tag{71}$$

which plots as in Figure 22. When multiple binding sites are present on the same receptor the Scatchard plot will assume a convex curvature.

IV. DISCREPANCIES IN AND ALTERNATIVES FOR THE HILL EQUATION

Equation 70 was derived under the assumption that there are many cooperating binding sites for the same ligand on a receptor. If this assumption holds true, the value obtained for n from the Hill plots will be the number of cooperating binding sites on each receptor molecule. This number should be a simple integer. It is more often the case, however, that the n value obtained is less than the number of agonist binding sites on the receptor. It also often occurs that the Hill plot of actual experimental data is nonlinear.

After taking experimental error into account, these discrepancies suggest that some of the assumptions used in the derivation of the Hill equation may not hold accurately when applied to actual receptor data. Since the number of binding sites on a receptor molecule is generally small, it is not difficult to derive a set of equations for actual numbers of binding sites. Then, depending on the magnitude of the n value obtained from the Hill analysis, experimental data can be fit to the equations, representing discrete numbers of binding sites. Furthermore, employing equations derived for discrete numbers of binding sites allows the flexibility of analyzing for the contribution that each site has on the effect produced by the ligand. This also allows an assessment of the relative magnitudes of the affinities of the individual ligand binding sites.

If we begin with a receptor that has two binding sites

$$R + L \rightleftharpoons RL + L \rightleftharpoons RLL \qquad \text{(Mechanism 8)}$$

where

$$[R_T] = [R] + [RL] + [RLL] \qquad (72)$$

$$K_{D_1} = \frac{[R]\,[L]}{[RL]} \qquad (73)$$

and

$$K_{D_2} = \frac{[RL]\,[L]}{[RLL]} \qquad (74)$$

Then, when only RLL is capable of producing an effect, we have

$$\frac{E_L}{E_{Lmax}} = \frac{[RLL]}{[R] + [RL] + [RLL]} \qquad (75)$$

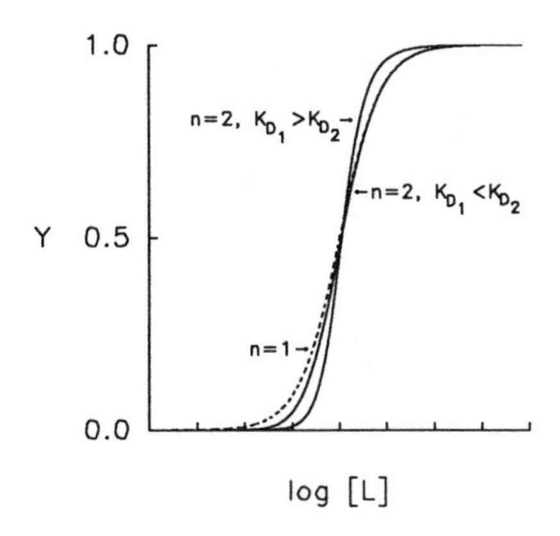

FIGURE 23

Substituting from Equations 73 and 74 into Equation 75 and rearranging yields

$$\frac{E_L}{E_{Lmax}} = \frac{1}{\frac{K_{D_2}}{[L]}\left(1 + \frac{K_{D_1}}{[L]}\right) + 1} \tag{76}$$

Plotting Y vs. log [L] from Equation 76 yields Figure 23. Figure 23 is not noticeably different from Figure 20. However, when we rearrange Equation 76 to its Hill equation equivalent form we have

$$\log\left(\frac{Y}{1\text{-}Y}\right) = \log [L] - \log K_{D_2} - \log\left(1 + \frac{k_{D_1}}{[L]}\right) \tag{77}$$

Equation 77 is not an equation for a straight line. When [L] is very large compared with K_{D_1} the log (1 + K_{D_1}/[L]) term becomes negligibly small and Equation 77 simplifies to Equation 70 where n = 1. Similarly, when [L] is very small compared with K_{D_1} the 1 in the log (1 + K_{D_1}/[L]) term becomes insignificant and Equation 77 simplifies to

$$\log\left(\frac{Y}{1\text{-}Y}\right) = 2 \log [L] - \log\left(K_{D_1}K_{D_2}\right) \tag{78}$$

Equation 78 is a form of Equation 70 where n = 2. The results are the same whether $K_{D_1} > K_{D_2}$, $K_{D_1} = K_{D_2}$, or $K_{D_1} < K_{D_2}$. Plotting log (Y/1-Y) vs. log [L] from Equation 77 yields Figure 24. Here we have a curved line that approaches a slope of 2 at low [L] and a slope of 1 at high [L].

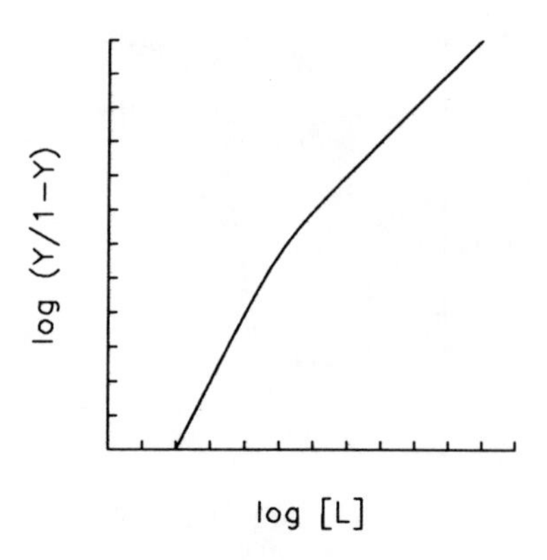

FIGURE 24

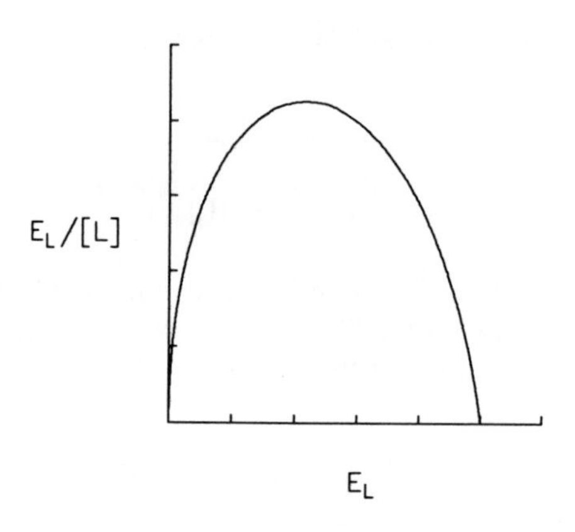

FIGURE 25

Rearranging Equation 75 to its Scatchard equivalent form yields

$$\frac{E_L}{[L]} = -\frac{E_L}{K_{D_2}\left(1 + \frac{K_{D_1}}{[L]}\right)} + \frac{E_{Lmax}}{K_{D_2}\left(1 + \frac{K_{D_1}}{[L]}\right)} \tag{79}$$

Equation 79 plots as shown in Figure 25.

When both RL and RLL from Mechanism 8 are capable of producing a physiological effect we have

$$\frac{E_L}{E_{Lmax}} = \frac{[RL] + [RLL]}{[R] + [RL] + [RLL]} \tag{80}$$

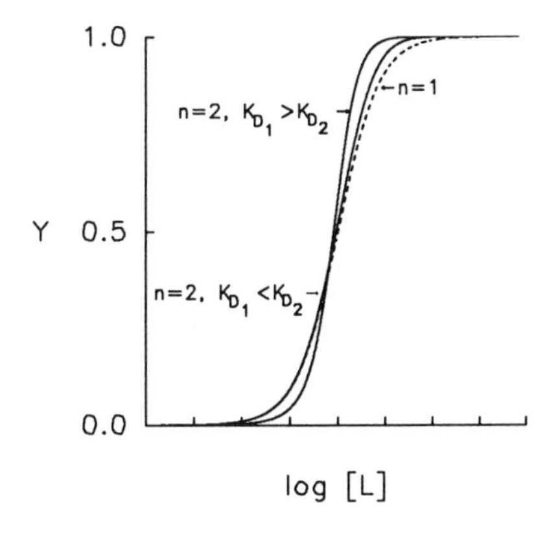

FIGURE 26

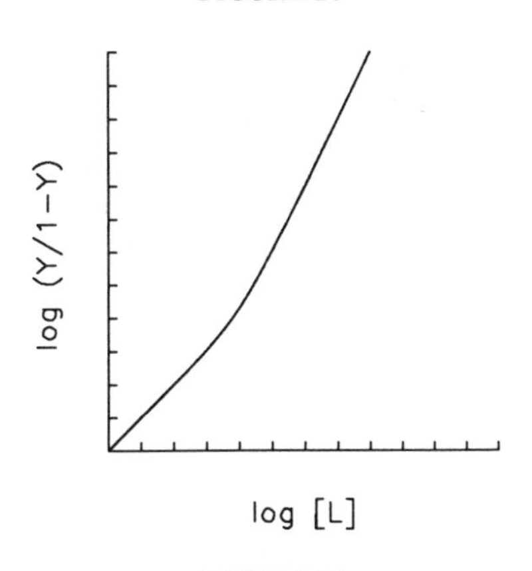

FIGURE 27

Substituting from Equations 73 and 74 into Equation 80 and rearranging yields

$$\frac{E_L}{E_{Lmax}} = \frac{\left(1 + \frac{K_{D_2}}{[L]}\right)}{\frac{K_{D_2}}{[L]}\left(1 + \frac{K_{D_1}}{[L]}\right) + 1} \tag{81}$$

Plotting Y vs. log [L] from Equation 81 yields Figure 26. Again, Figure 26 is not noticeably different from Figure 20. Rearranging Equation 78 to its Hill equation equivalent form yields

$$\log\left(\frac{Y}{1\text{-}Y}\right) = 2 \log [L] - \log\left(K_{D_1}K_{D_2}\right) + \log\left(1 + \frac{K_{D_2}}{[L]}\right) \tag{82}$$

Plotting log (Y/1-Y) vs. log [L] from Equation 82 yields Figure 27. Again,

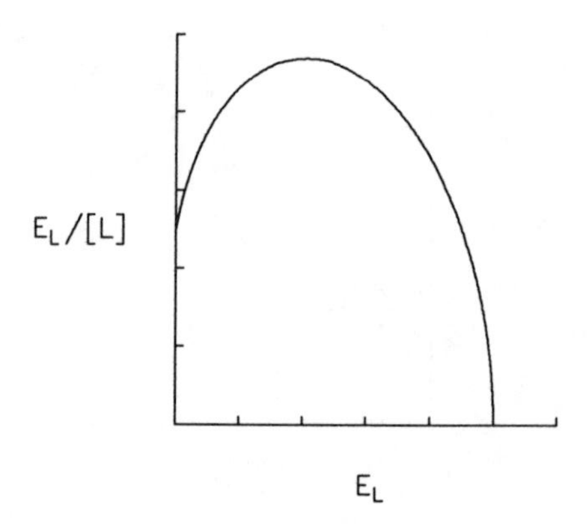

FIGURE 28

we see that a plot of log (Y/1-Y) vs. log [L] from Equation 82 does not give a straight line. When [L] is large compared with K_{D_2} the log $(1 + K_{D_2}/[L])$ term becomes negligibly small and Equation 82 simplifies to Equation 78 where $n = 2$. When [L] is small compared with K_{D_2} the 1 in the $(1 + K_{D_2}/[L])$ terms becomes negligible and Equation 82 simplifies to

$$\log\left(\frac{Y}{1-Y}\right) = \log [L] - \log K_{D_1} \tag{83}$$

where $n = 1$. Again, the results are the same whether $K_{D_1} > K_{D_2}$, $K_{D_1} = K_{D_2}$, or $K_{D_1} < K_{D_2}$.

Equation 81 can also be rearranged to its Scatchard equivalent form

$$\frac{E_L}{[L]} = -\frac{E_L}{K_{D_2}\left(1 + \frac{K_{D_1}}{[L]}\right)} + \frac{E_{Lmax}\left(1 + \frac{K_{D_2}}{[L]}\right)}{K_{D_2}\left(1 + \frac{K_{D_1}}{[L]}\right)} \tag{84}$$

which plots as shown in Figure 28. The curve in Figure 28 differs from those in Figures 22 and 25 by intersecting the ordinate above zero instead of at zero.

V. PHYSIOLOGICAL ADVANTAGES OF HOMOTROPIC COOPERATIVITY

What are the physiological advantages of homotropic cooperativity? Returning to a plot of Y vs. log [L] shown in Figure 20 we see that the sigmoid curve becomes steeper when we have cooperativity. What this means physiologically is that receptors with multiple ligand binding sites have narrower effective concentration ranges than those with only one ligand binding site.

A. HEMOGLOBIN

The physiological importance of this narrowing of the effective concentration range can be illustrated by the blood protein, hemoglobin, and its binding of oxygen. Hemoglobin functions to transport oxygen, by way of the bloodstream, from the lungs to all the tissues of the body. Each hemoglobin molecule has four oxygen binding sites. When hemoglobin enters the lungs, where the oxygen concentration is high, all the oxygen binding sites become occupied with oxygen over a very narrow concentration range. This ensures that the hemoglobin molecule becomes saturated with oxygen during its brief passage through the lungs. Even more importantly, it provides for a much more uniform distribution of oxygen throughout the body.

As soon as the hemoglobin leaves the lungs the concentration of free oxygen decreases dramatically. This means that we are moving in the direction of decreasing fractional receptor occupancy. If there was no cooperativity in the binding of oxygen to hemoglobin, a very small concentration drop in free oxygen upon leaving the lungs would result in substantial amounts of oxygen dissociating from the hemoglobin. As the blood moves farther away from the lungs the oxygen concentration would become still lower and still more of the bound oxygen would dissociate. By the time the blood reached the toes all the oxygen would be gone and the toes would never get any. Since there are four oxygen binding sites on each hemoglobin molecule, the Y vs. log [L] curve is even steeper than the one shown in Figure 20. With the steepened sigmoid curve, the oxygen concentration can drop much further before any significant amounts of oxygen dissociate from the hemoglobin. Then all the oxygen will dissociate over a much narrower concentration range. In such a way, each hemoglobin molecule travels much farther from the lungs before discharging its oxygen. Then instead of discharging the oxygen gradually, the hemoglobin molecules discharge their oxygen nearly all at once when they reach an area where the oxygen concentration is low enough. Thus, the distribution of oxygen throughout the body is much more uniform than would be possible without cooperative binding.

B. CALMODULIN

Calmodulin is a cytoplasmic protein that functions to bind calcium. The calcium-calmodulin complex, in turn, binds to various enzymes and proteins in the cell to alter their activities, thereby producing the physiological effect. Each calmodulin molecule has four binding sites for calcium, which, like hemoglobin, exhibit cooperativity. In the calcium-calmodulin system the calcium enters the cell cytoplasm from outside the cell through calcium channels or it is released from stores in the endoplasmic reticulum. The calmodulin, in its role in signal transduction, needs to be able to respond in a nearly all-or-none fashion to calcium concentration changes. It must go from a near zero binding site occupancy to near fully occupied binding sites over a very narrow concentration range when calcium enters the cytoplasm. The calmodulin must then go from near fully occupied to near fully unoccupied over the same very narrow concentration range as the calcium is pumped out of the cytoplasm. The cooperativity of its interaction with calcium permits calmodulin to function in the manner required.

VI. HETEROTROPIC COOPERATIVITY

To this point we have only discussed the circumstances of multiple binding sites for the same ligand on a receptor. We refer to the interactions of a single receptor with multiple binding sites for the same ligand as homotropic. Receptor function also may be modulated by inhibitors or stimulators, which are ligands other than the agonist and which bind at sites other than the agonist binding site. These modulatory ligands may cause the affinity of the agonist binding site to be altered when they bind. Interaction between binding sites for two or more nonidentical ligands on a receptor, which mutually influence the affinities for one another, is heterotropic cooperativity. Heterotropic cooperativity can, under certain circumstances, cause disproportionalities between receptor occupancy and receptor stimulation. Heterotropic cooperativity will be addressed further in Chapter 8.

VII. RECEPTOR DESENSITIZATION

The fourth circumstance that can cause a disproportionality between fractional receptor occupancy and physiological effect occurs when receptors can undergo a change to a desensitized form. Desensitization can be represented by Mechanism 9.

$$R + L \rightleftharpoons RL_S \rightleftharpoons RL_{des} \qquad \text{(Mechanism 9)}$$

RL_S is the stimulated receptor-ligand complex and RL_{des} is the desensitized receptor-ligand complex. The nicotinic acetyl choline receptor of the mammalian neuromuscular junction is an example of a receptor system that automatically desensitizes shortly after the stimulated receptor-ligand complex has formed. Desensitization functions physiologically to provide an automatic off switch for this receptor-mediated signal transduction system. Often, desensitization is accompanied by a decrease in the affinity of the receptor for its agonists, but this is not always the case. With desensitization, B_L is not proportional to S_L. If desensitization changes the affinity for L it could cause Eadie-Hofstee or Rosenthal-Scatchard plots to be nonlinear. This is shown in Figure 29. The concave curvature of Figure 29 is the result of a summation of binding to two sets of sites having different affinities and different $B_{L\max}$ values. This is indicated by the dashed lines. This behavior is described by Equation 85:

$$\frac{B_L}{[L]} = -\frac{B_L}{K_{D_1}} + \frac{B_{L\max_1}}{K_{D_1}} - \frac{B_L}{K_{D_2}} + \frac{B_{L\max_2}}{K_{D_2}} \qquad (85)$$

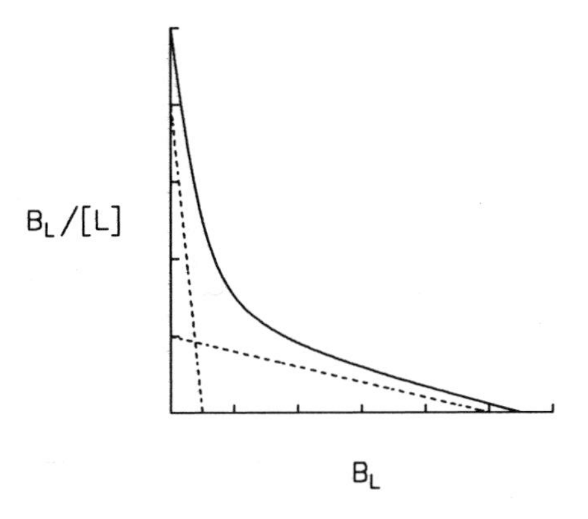

FIGURE 29

For receptor desensitization one set of K_D and B_{Lmax} values is contributed by RL_S and the other set is contributed by RL_{des}. Anytime two or more receptor sites with different affinities **and** different B_{Lmax} values combine to produce an effect or to produce total binding, curved Rosenthal-Scatchard or Eadie-Hofstee plots can result. In order for the curves to be concave the receptor with the low affinity must be the one with the high B_{Lmax}. The reverse can produce a convex curvature.

Chapter 8

ANTAGONISM

I. INTRODUCTION

Antagonists are substances that inhibit receptor stimulation by agonists or, they are substances that inhibit the effects of receptor stimulation by agonists. A simplistic way of analyzing inhibition is to generate an effect vs. log [I] curve as shown in Figure 30. Figure 30 can be generated experimentally by holding the agonist concentration constant at some level that produces a measurable effect less than E_{Lmax} and varying the concentration of I. The abscissal midpoint of the curve is the IC_{50} (inhibitor concentration that gives 50% of the uninhibited effect). Inhibition curves generated in this manner can have the same shape (although going in the opposite direction) as simple effect vs. log [L] plots. Under a variety of conditions, including multiple cooperative inhibitor binding sites, they may be steeper. This is shown in Figure 31. Without additional information such analyses cannot provide information on the mechanism of inhibition. Such data are also incapable of providing the researcher with any indication of the relationship between the IC_{50} value and the affinity of the receptor for the inhibitor. The only value such analyses have is for providing relative ordering of inhibition potencies for a series of inhibitors.

II. TYPES OF ANTAGONISM

There are several types of antagonism. These are listed below according to the mechanism of antagonism.

A. CHEMICAL ANTAGONISM

In chemical antagonism the antagonist directly interacts with the agonist to render the agonist inactive. There is no involvement of the receptor in this type of antagonism. Chemical antagonism has the effect of decreasing [L]. An enzyme that degrades the agonist or a different receptor that binds the agonist would both fit the definition of a chemical antagonist. This type of antagonism occurs often and cannot be ignored, but it won't be covered further in this book.

B. PHYSICAL ANTAGONISM

When two independent agonists stimulate two independent receptor systems to cause counteracting effects on two independent effector systems we have physical antagonism. Following is an example of physical antagonism: agonist 1 stimulates receptors in the heart to produce an increased force of

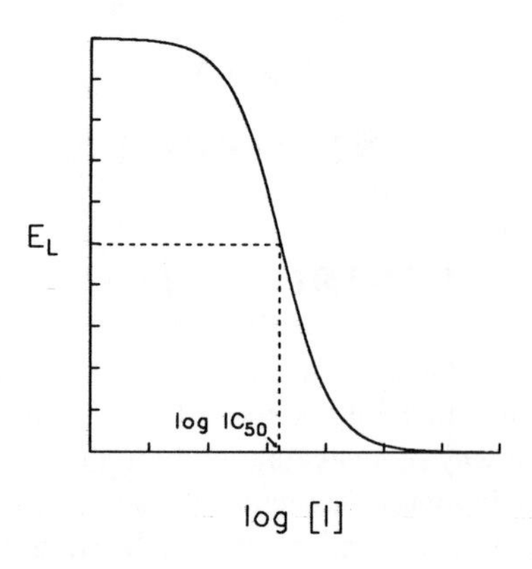

FIGURE 30

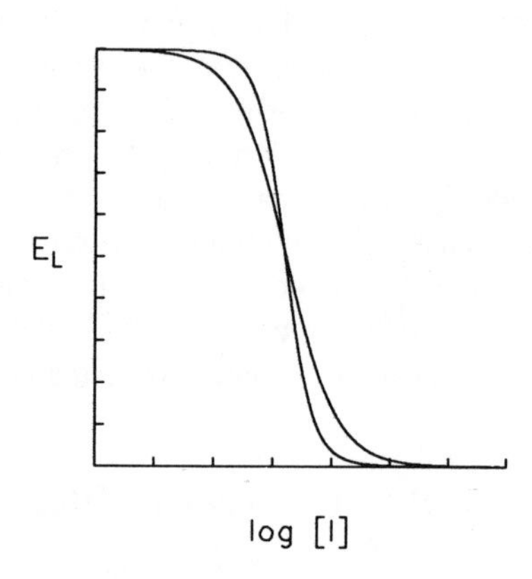

FIGURE 31

contraction; this has the effect of increasing the blood pressure. Agonist 2 stimulates receptors in the peripheral vasculature to cause vasodilation. This has the effect of decreasing blood pressure. Counteracting receptor-effector systems, such as the one described, are the rule rather than the exception in physiology and are very important in maintaining homeostasis. We will not discuss physical antagonism further.

C. FUNCTIONAL ANTAGONISM

Functional antagonism is similar to physical antagonism. Two independent agonists stimulate two independent receptor systems to produce opposing effects on the *same* effector system. An example of functional antagonism is the G protein-regulated adenylate cyclase system. Adenylate cyclase is an enzyme associated with the cytoplasmic side of most cell membranes. This enzyme functions to convert ATP to cyclic AMP. Cyclic AMP is a substance that activates a variety of protein kinases inside the cell. These, in turn, phosphorylate key cellular enzymes and other proteins, thereby altering their activities. A large number of agonists and their receptors operate through G proteins to influence the activity of adenylate cyclase. One example would be a β adrenergic receptor linked through a G_S (stimulatory) protein that activates adenylate cyclase and, in the same membrane, a D_2 dopamine receptor linked through a G_I (inhibitory) protein that inhibits the same adenylate cyclase. Generally, the inhibitory event will override the stimulatory event when both occur together. As was true for physical antagonism, functional antagonism is widespread in physiological systems and is important for maintaining cellular homeostasis. Functional antagonism will not be discussed further.

D. IRREVERSIBLE ANTAGONISTS

Irreversible antagonists are substances that interact with the receptor to inhibit it permanently. These have the effect of reducing $[R_T]$. Irreversible antagonists generally act by undergoing a chemical reaction with the receptor. Enzymatic phosphorylation of receptors qualifies as a form of irreversible antagonism. This is true even though there are generally phosphatase enzymes available to dephosphorylate the receptor when conditions require its reactivation. Researchers have long employed ''suicide'' ligands to tag receptor sites chemically. These are compounds that have affinity and specificity for the receptor site in question. Once bound they can be activated, such as by exposure to light, to undergo a chemical reaction with the receptor and become chemically bonded to it. We will not discuss irreversible antagonists further.

E. REVERSIBLE ANTAGONISTS

Antagonists in this classification are those that interact directly with the receptor or with an effector system functionally coupled to the receptor. Reversible antagonists can be subclassified into six types; simple competitive, allosteric competitive, simple noncompetitive, heterotropic-cooperative noncompetitive, homotropic noncompetitive, and uncompetitive. The remainder of this chapter will address these forms of antagonism, and corresponding stimulatory mechanisms.

1. Simple Competitive Antagonism

A simple competitive antagonist is a ligand that binds to the same site on a receptor as the agonist, but is incapable of stimulating the receptor. Simple competitive antagonists have zero intrinsic activity. When the antagonist binds in the agonist binding site it produces antagonism by occupying the site and preventing the agonist from binding and stimulating the receptor. Returning to the lock and key analogy, if you insert the wrong key into a lock it will not unlock it. Your error has delayed (or inhibited) your entry into your house. Thus, the antagonist and the agonist compete with one another

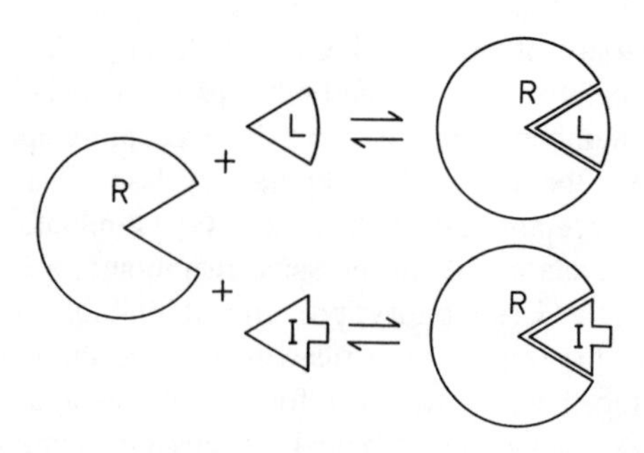

FIGURE 32

for access to the same binding site. Mechanism 10 and Figure 32 describe simple competitive antagonism.

$$\begin{array}{c} R + L \underset{k_{-1}}{\overset{k_{+1}}{\rightleftharpoons}} RL \\ + \\ I \\ k_{+2} \uparrow \downarrow k_{-2} \\ RI \end{array} \qquad \text{(Mechanism 10)}$$

The symbol I is used to designate the antagonist (inhibitor) and to differentiate it from the agonist (L). This mechanism shows that the unoccupied receptor can combine with either the agonist (L) or the antagonist (I) to form either an RL complex or an RI complex. Since both of these reactions are reversible, either complex can dissociate to regenerate the unoccupied recep-26tor. The unoccupied receptor can then form a complex with either L or I.

We need equations that include the antagonist such that we can analyze its effects on the receptor system.

Recalling Equation 8 we have

$$K_D = \frac{k_{-1}}{k_{+1}} = \frac{[R]\,[L]}{[RL]} \tag{8}$$

and similarly,

$$K_I = \frac{k_{-2}}{k_{+2}} = \frac{[R]\,[I]}{[RI]} \tag{86}$$

K_I is the dissociation constant for the receptor inhibitor complex. Rearranging Equations 8 and 86 we obtain

$$[RL] = \frac{[R]\,[L]}{K_D} \tag{87}$$

and

$$[RI] = \frac{[R]\,[I]}{K_I} \tag{88}$$

We know that

$$[R_T] = [R] + [RL] + [RI] \tag{89}$$

and

$$\frac{B_L}{B_{Lmax}} = \frac{[RL]}{[R_T]} = \frac{[RL]}{[R] + [RL] + [RI]} \tag{90}$$

Substituting for [RL] and [RI] gives

$$\frac{B_L}{B_{Lmax}} = \frac{\dfrac{[R]\,[L]}{K_D}}{[R] + \dfrac{[R]\,[L]}{K_D} + \dfrac{[R]\,[I]}{K_I}} \tag{91}$$

Rearranging gives

$$\frac{B_{Lmax}}{B_L} = \frac{[R] + \dfrac{[R]\,[L]}{K_D} + \dfrac{[R]\,[I]}{K_I}}{\dfrac{[R]\,[L]}{K_D}} \tag{92}$$

Expanding and canceling terms yields

$$\frac{B_{Lmax}}{B_L} = \frac{K_D}{[L]}\left(1 + \frac{[I]}{K_I}\right) + 1 \tag{93}$$

Further rearrangement gives

$$B_L = \frac{B_{Lmax}\,[L]}{K_D\left(1 + \dfrac{[I]}{K_I}\right) + [L]} \tag{94}$$

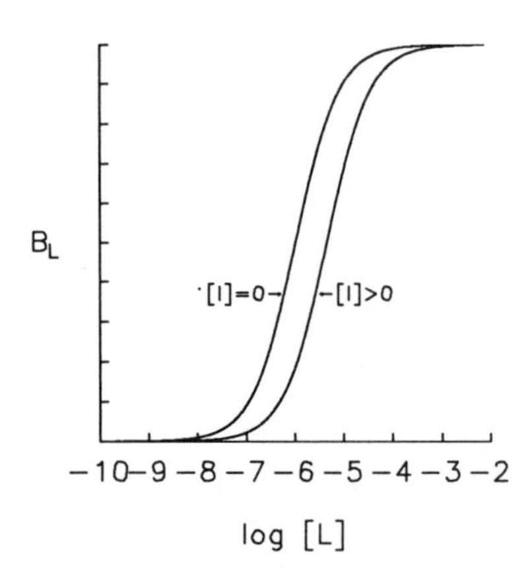

FIGURE 33

Equation 94 is in the same form as Equation 18:

$$B_L = \frac{B_{Lmax}\,[L]}{K_D + [L]} \tag{18}$$

except for the $(1 + [I]/K_I)$ term for the inhibition. Note that if $[I] = 0$, this term will drop out of Equation 94, giving Equation 18. Graphically, competitive inhibitors cause parallel shifts of the B_L vs. log [L] curve to the right without altering B_{Lmax}. This is shown in Figure 33.

How can we have inhibition without altering B_{Lmax}? Consider a circumstance where you have a key ring with two keys on it. Each key is cut from the same blank so each of them will fit into the lock but only one of them will unlock it. It is dark and you cannot see the keys. In order to get into your house, to prepare a meal, to satisfy your hunger, you have to get the right key into the lock and unlock it. The odds are 50% that you will choose the wrong key on your first try. The extra key thus serves to inhibit your entry into the house. Now suppose that you have 20 keys on the ring. Each of them will fit into the lock but only one of them will unlock it. The odds are that you will have to try many keys before you find the right one. Therefore, it takes you much longer to gain entry. This illustrates that the degree of inhibition is dependent on the relative concentration ratio of agonist to antagonist. Now reverse the circumstances and place 19 keys on the ring that will unlock the lock and only one that will not. The odds are that you will choose a correct key almost every try. This illustrates that you can nearly eliminate inhibition by changing the relative concentration ratio in favor of the agonist. When the agonist concentration approaches infinity, this ratio also approaches infinity and the inhibition is nearly eliminated. **Competitive inhibitors do not alter B_{Lmax}. They also do not alter K_D. The value of K_D is the same in Equation 94 as it is in Equation 18.** Since the antagonist

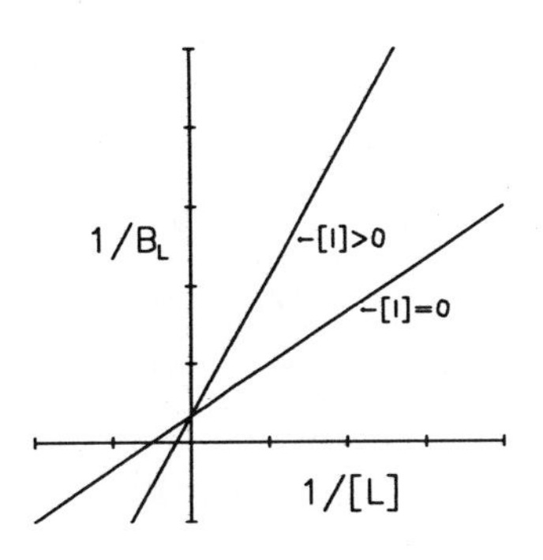

FIGURE 34

and the agonist cannot be bound to the receptor simultaneously, there is no way that the antagonist can influence affinity for the agonist.

Figure 33 shows us that for any given level of B_L the presence of inhibitor requires a higher concentration of L. The same effect can be seen when we transform Equation 94 to a linear form such as the double-reciprocal equation for simple competitive inhibition,

$$\frac{1}{B_L} = \frac{1}{[L]} \frac{K_D}{B_{Lmax}} \left(1 + \frac{[I]}{K_I}\right) + \frac{1}{B_{Lmax}} \tag{95}$$

which plots as shown in Figure 34. Here we see that the inhibitor affects only the slope term of the equation and not the $1/B_L$ axis intercept. The $1/[L]$ axis intercept for the inhibited line is equal to $-1/K_D(1 + [I]/K_I)$. The K_I can be evaluated from the slope of Figure 34 as follows:

$$\text{Slope}_L = \frac{K_D}{B_{Lmax}} \tag{96}$$

and

$$\text{Slope}_{L(I)} = \frac{K_D}{B_{Lmax}} \left(1 + \frac{[I]}{K_I}\right) \tag{97}$$

Substituting for K_D/B_{Lmax} yields

$$\text{Slope}_{L(I)} = \text{Slope}_L \left(1 + \frac{[I]}{K_I}\right) \tag{98}$$

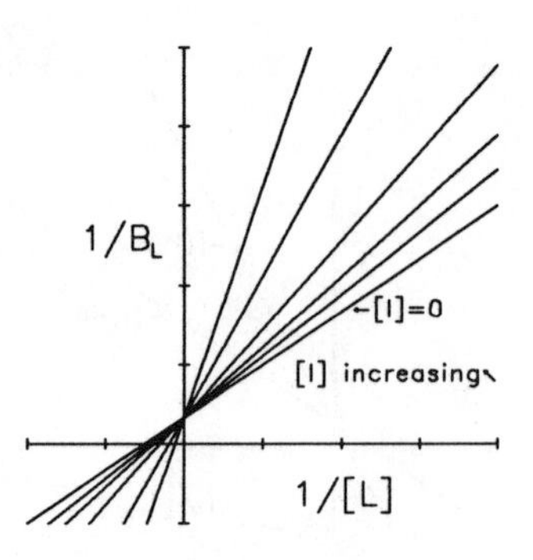

FIGURE 35

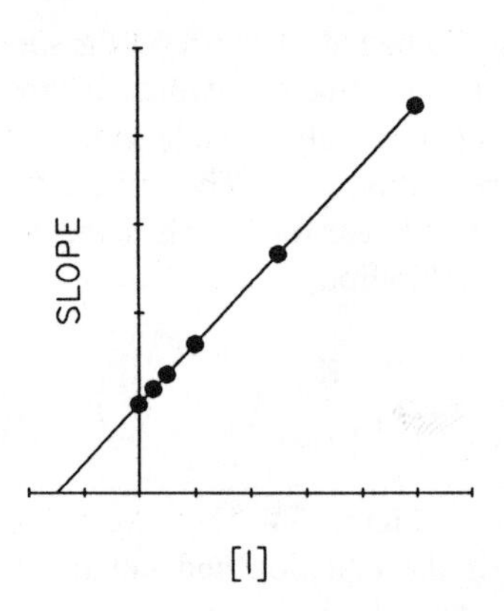

FIGURE 36

Rearranging gives

$$K_I = \frac{[I]}{\frac{Slope_{L(I)}}{Slope_L} - 1} \quad (99)$$

a. Slope Replots Of Double-Reciprocal Data

Slope replots of double-reciprocal data can also be used to evaluate the K_I. These have the additional capability of revealing allosteric competitive inhibition. For slope replot analysis the researcher generates a family of double-reciprocal graphs by analyzing several different inhibitor concentrations. This is shown in Figure 35. Then, the slopes of the double-reciprocal lines are replotted vs. [I] as shown in Figure 36. The slope of the slope replot is equal to $K_D/(K_I B_{Lmax})$. A linear slope replot indicates a simple competitive mechanism of inhibition. However, as discussed later in this chapter, some

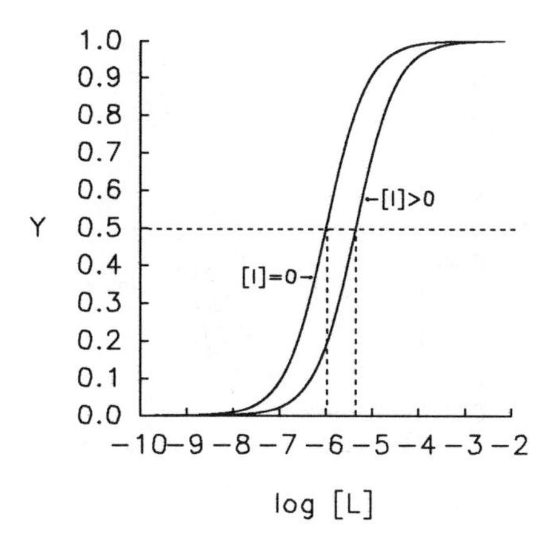

FIGURE 37

forms of heterotropic-cooperative noncompetitive inhibition can produce double-reciprocal lines that intersect on the ordinate and produce linear slope replots. Allosteric competitive inhibition, discussed later in this chapter, will produce curved slope replots.

***b.** The Schild Equation*

Alternatively, the Schild relationship can be used to evaluate competitive antagonism.[12] If we draw a horizontal line anywhere through Figure 33, (for convenience let us draw it through $Y = 0.5$) we get Figure 37. Then the ligand concentration where it intersects the uninhibited receptor will be [L] and the ligand concentration where it intersects the inhibited receptor will be [L′]. Then, since the two ligand concentrations give the same Y value, by combining Equations 18 and 94 we arrive at the following relationship:

$$\frac{[L]}{K_D + [L]} = \frac{[L']}{K_D\left(1 + \frac{[I]}{K_I}\right) + [L']} \tag{100}$$

Rearranging yields

$$\frac{K_D}{[L]} + 1 = \frac{K_D\left(1 + \frac{[I]}{K_I}\right)}{[L']} + 1 \tag{101}$$

and then

$$\frac{[L']}{[L]} = 1 + \frac{[I]}{K_I} \text{ or } \frac{[L']}{[L]} - 1 = \frac{[I]}{K_I} \tag{102}$$

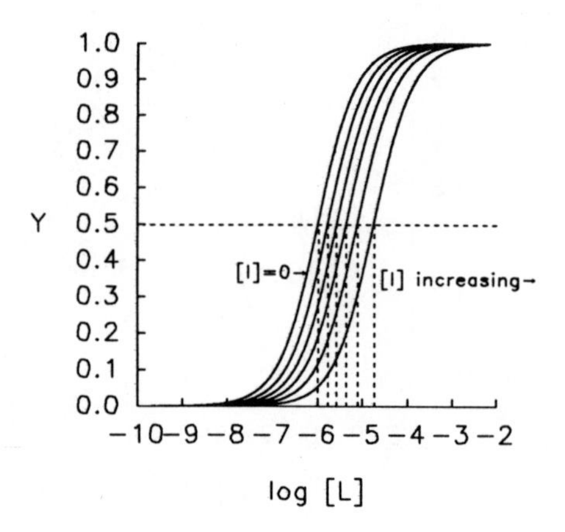

FIGURE 38

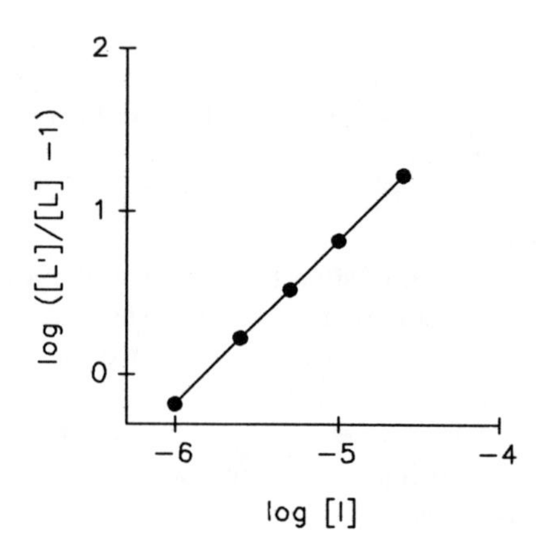

FIGURE 39

Taking the log, we get

$$\log\left(\frac{[L']}{[L]} - 1\right) = \log [I] - \log K_I = \log [I] + pK_I \qquad (103)$$

the Schild equation. Evaluation of the Schild equation requires that the researcher generate a family of Y vs. log [L] curves for a series of I concentrations, such as the one shown in Figure 38. This provides a set of [L′] values that can be plotted as a Schild plot, Figure 39.

If the Schild plot is linear with a slope of 1 this is an indication of simple competitive inhibition. However, as was true for slope replots, some forms of heterotropic-cooperative noncompetitive inhibition also can produce linear Schild plots. Allosteric competitive inhibition will produce curved Schild plots. In a more general situation, where the ratio of ligand to inhibitor is not 1:1, *e.g.*, more than one inhibitor molecule is required to bind to each receptor to produce the inhibition, we have

$$\log\left(\frac{[L']}{[L]} - 1\right) = n \log [I] + pK_I \qquad (104)$$

where n is the slope of the Schild plot and is equal to the number of inhibitor molecules required to bind to each receptor to produce the inhibition.

c. The Value pI_2

The Schild relationship also can be employed to determine another value, the pI_2. The pI_2 is more commonly known in the literature as the pA_2. pI_2 is the $-\log$ of the simple competitive inhibitor concentration that requires an exact doubling of the agonist concentration from the uninhibited system to produce the same effect. There is a deviation here from the standard literature designation in the interest of maintaining consistency in terminology and to reduce possible ambiguity in the use of terms. There are many values that could be referred to as *A*. These include agonist, antagonist, affinity, and activator. We have already used *A* or *a* three times for Association, *a*cid, and *a*ctivity. Therefore, *I* is used for inhibitor instead of *A* for antagonist throughout this book. Substituting for $[L'] = 2[L]$ into Equation 103 gives

$$\log\left(\frac{2\,[L]}{[L]} - 1\right) = \log 1 = 0 = \log [I] + pK_I \qquad (105)$$

or

$$-\log [I] = pK_I \text{ or } pI_2 = pK_I \qquad (106)$$

If the concentration of I at the receptors is known, and when the conditions of B_L proportional to S_L and S_L proportional to E_L hold, the pI_2 is equal to the pK_I. If these conditions do not hold then the pI_2 measured from effect data may not be equal to the pK_I.

2. Partial Agonists As Competitive Inhibitors Of Full Agonists

Partial agonists, those where $0 < \alpha < 1$, can function as partial competitive antagonists of full agonists. We have

$$\begin{array}{l} R + L_1 \rightleftharpoons RL_1 \\ + \\ L_2 \\ \uparrow\downarrow \\ RL_2 \end{array} \qquad \text{(Mechanism 11)}$$

where

$$K_{D_1} = \frac{[R]\,[L_1]}{[RL_1]} \tag{107}$$

$$K_{D_2} = \frac{[R]\,[L_2]}{[RL_2]} \tag{108}$$

$$[R_T] = [R] + [RL_1] + [RL_2] \tag{109}$$

and

$$E_{L_1L_2} = E_{L_1(L_2)} + E_{L_2(L_1)} \tag{110}$$

$E_{L_1L_2}$ is the effect of L_1 plus L_2. $E_{L_1(L_2)}$ is the effect of L_1 in the presence of L_2 and $E_{L_2(L_1)}$ is the effect of L_2 in the presence of L_1. Then

$$\frac{E_{L_1(L_2)}}{E_{max}} = \frac{\alpha_1}{\frac{K_{D_1}}{[L_1]}\left(1 + \frac{[L_2]}{K_{D_2}}\right) + 1} \tag{111}$$

$$\frac{E_{L_2(L_1)}}{E_{max}} = \frac{\alpha_2}{\frac{K_{D_2}}{[L_2]}\left(1 + \frac{[L_1]}{K_{D_1}}\right) + 1} \tag{112}$$

and

$$\frac{E_{L_1L_2}}{E_{max}} = \frac{\alpha_1}{\frac{K_{D_1}}{[L_1]}\left(1 + \frac{[L_2]}{K_{D_2}}\right) + 1} + \frac{\alpha_2}{\frac{K_{D_2}}{[L_2]}\left(1 + \frac{[L_1]}{K_{D_1}}\right) + 1} \tag{113}$$

Note that if either $\alpha = 0$ that term drops out of the equation. If the remaining α is 1 we have the equation for simple competitive inhibition. Plotting $E_{L_1L_2}/E_{max}$ vs. log $([L_1]/K_{D_1})$ for increasing concentrations of L_2 where α_1 and α_2 are both equal to 1 results in the family of curves shown in Figure 40. From Figure 40 we see that as the starting concentration of L_2 increases we start at higher levels on the ordinate and produce a series of sigmoid curves which all approach a limiting value of 1 on the ordinate scale. Furthermore, the midpoints of all the sigmoid curves (where $[L_1]$ equals K_{D_1}) lie at zero on the abscissa.

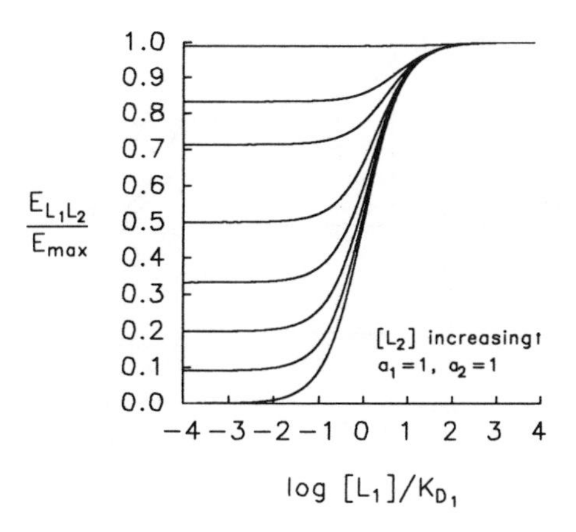

FIGURE 40

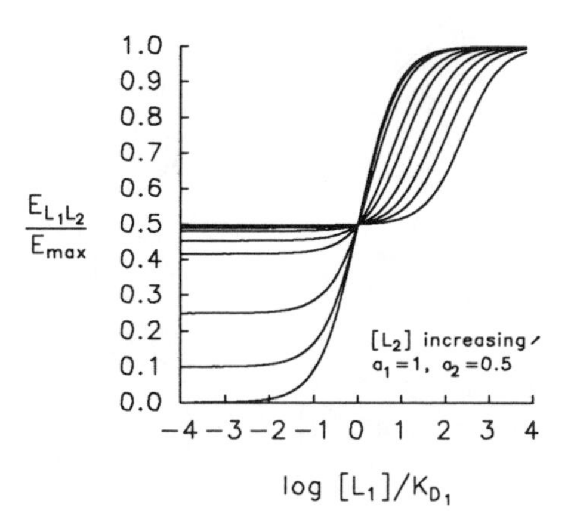

FIGURE 41

The same plot where α_1 equals 1 and α_2 equals 0.5 results in Figure 41. Here the maximum possible starting ordinate value is 0.5. Moving to the right toward zero on the abscissa traverses a region where the effects of the two agonists are additive. All the curves cross one another at zero on the abscissa scale and at α_2 on the ordinate scale. For abscissa values of greater than zero, we see a series of sigmoid curves for which the midpoints progressively move to the right as the concentration of L_2 increases. In the lower left-hand quadrant of the graph the effects of the two agonists are additive. In the upper right-hand quadrant of the graph the presence of L_2 is inhibiting L_1. When α_1 equals 0.5 and α_2 equals 1 we get Figure 42. In the lower half of Figure 42 we see the additive effects of the two ligands and the curves approach α_1 on the ordinate scale as their maximum value. In the upper half of the graph we see inhibition of L_2 by L_1 in the form of a series of downward-going sigmoid

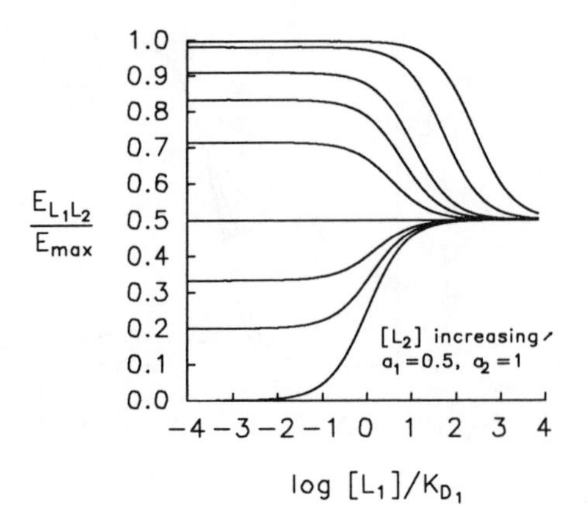

FIGURE 42

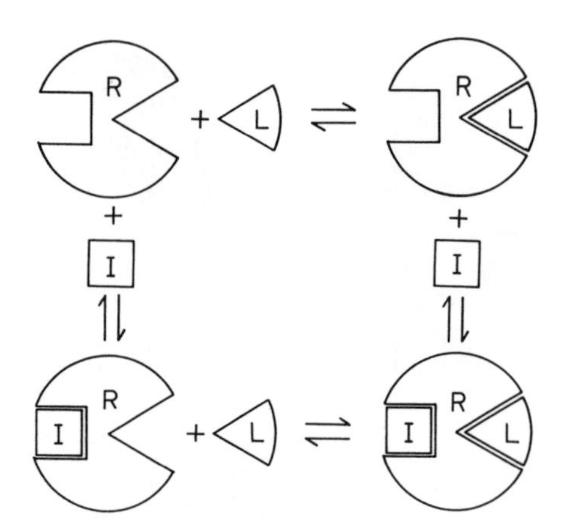

FIGURE 43

curves that approach α_1 on the ordinate scale as their minimum value. In addition, the midpoints of the sigmoid curves in the upper half of the graph move progressively to the right as the concentration of L_2 increases.

3. Simple Noncompetitive Antagonism

Simple noncompetitive antagonism can be described by Mechanism 12 and Figure 43.

$$\begin{array}{ccc} R + L & \rightleftharpoons & RL \\ + & & + \\ I & & I \\ \uparrow\downarrow & & \uparrow\downarrow \\ RI + L & \rightleftharpoons & RLI \end{array} \qquad \text{(Mechanism 12)}$$

In this mechanism the antagonist binds to a site on the receptor that is separate and independent of the agonist binding site. Mathematically stated

$$K_{D_1} = \frac{[R]\,[L]}{[RL]} = K_{D_2} = \frac{[RI]\,[L]}{[RLI]} \tag{114}$$

and,

$$K_{I_1} = \frac{[R]\,[I]}{[RI]} = K_{I_2} = \frac{[RL]\,[I]}{[RLI]} \tag{115}$$

With simple noncompetitive antagonism, binding of the inhibitor does not influence the affinity of the receptor for the agonist. Similarly, binding of the agonist does not influence the affinity of the receptor for the inhibitor. Thus, the inhibitor can bind equally well to the unoccupied receptor or to the RL complex and the agonist can bind equally well to the unoccupied receptor or to the RI complex.

Measurements of binding only, which do not measure receptor stimulation or an effect due to receptor stimulation, will not detect simple noncompetitive inhibition. The reason for this is that B_L measures both RL and RLI and is incapable of distinguishing between them. When the affinity of the ligand is the same for the RI complex as it is for R, as stated in Equation 114, the equation for binding that describes Mechanism 12 simplifies to Equation 18. Whenever the inhibitor is bound, however, the receptor is either incapable of being stimulated by the agonist, or the stimulated receptor is incapable of producing its physiological effect. Therefore, measurements of physiological effect which distinguish between RL and RLI are required to allow us to measure the influence of the simple noncompetitive inhibitor.

A good example of a simple noncompetitive inhibitor would be a molecule that binds in the pore of a receptor-modulated ion channel and physically plugs the channel. When agonist binds to the receptor it stimulates the receptor, which, in turn, induces the channel to open. However, since the channel is plugged, no ions flow across the membrane and the physiological effect is blocked.

The equations for simple noncompetitive inhibition in terms of effect are

$$\frac{E_L}{E_{Lmax}} = \frac{1}{\dfrac{K_D}{[L]}\left(1 + \dfrac{[I]}{K_I}\right) + \left(1 + \dfrac{[I]}{K_I}\right)} \tag{116}$$

or

$$E_L = \frac{E_{Lmax}\,[L]}{K_D\left(1 + \dfrac{[I]}{K_I}\right) + [L]\left(1 + \dfrac{[I]}{K_I}\right)} \tag{117}$$

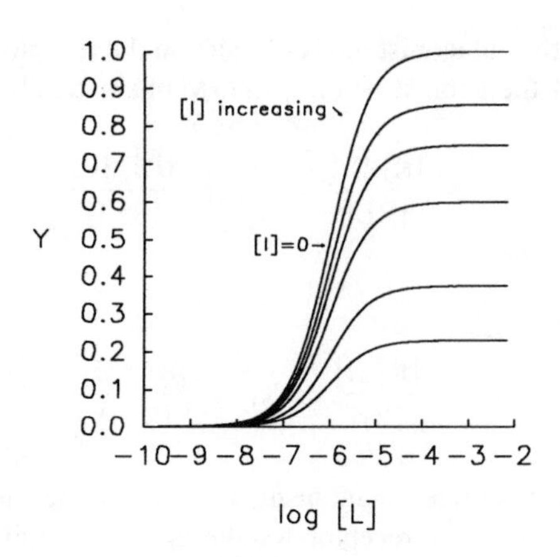

FIGURE 44

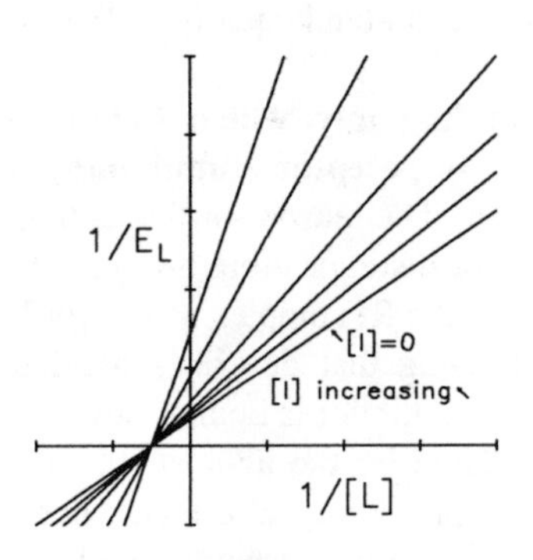

FIGURE 45

which plot as shown in Figure 44. Again, this is easier to see from the double-reciprocal form of the equation

$$\frac{1}{E_L} = \frac{1}{[L]}\frac{K_D}{E_{Lmax}}\left(1 + \frac{[I]}{K_I}\right) + \frac{1}{E_{Lmax}}\left(1 + \frac{[I]}{K_I}\right) \tag{118}$$

which graphs as shown in Figure 45. Here we see that the inhibitor does not change the abscissal midpoint of the sigmoid plot of Figure 44 (the affinity). It does, however, reduce the maximum effect possible for the receptor. From Figure 45, the inhibitor affects both the slope and the $1/E_L$ axis intercept but not the 1/[L] axis intercept.

Simple noncompetitive inhibitors reduce the maximum attainable effect due to agonist binding. They do so without influencing the concentration of the agonist that is required to attain any particular proportion of this new maximum effect. The reason for the reduction in maximum effect is that the inhibitor removes that proportion of receptor molecules that are complexed with it. This proportionately reduces E_{Lmax}. Increasing [L] cannot out-compete the inhibitor since L and I are binding to separate, independent sites.

4. Heterotropic-Cooperative Noncompetitive Inhibition

When inhibitor and agonist bind to separate but interacting sites **and**, when the binding of inhibitor prevents the agonist from stimulating the receptor **or**, when binding of the inhibitor prevents the stimulated receptor from producing an effect, we have heterotropic-cooperative noncompetitive inhibition. In this circumstance, for Mechanism 12, there will be two different K_D values and two different K_I values that describe the system, as follows:

$$K_{D_1} = \frac{[R]\,[L]}{[RL]} \tag{119}$$

$$K_{D_2} = \frac{[RI]\,[L]}{[RLI]} \tag{120}$$

$$K_{I_1} = \frac{[R]\,[I]}{[RI]} \tag{121}$$

and

$$K_{I_2} = \frac{[RL]\,[I]}{[RLI]} \tag{122}$$

In this circumstance

$$\frac{E_L}{E_{Lmax}} = \frac{[RL]}{[R] + [RL] + [RI] + [RLI]} \tag{123}$$

Substituting from Equations 119 to 122 into Equation 123 and rearranging yields

$$\frac{E_L}{E_{Lmax}} = \frac{[L]}{K_{D_1}\left(1 + \frac{[I]}{K_{I_1}}\right) + [L]\left(1 + \frac{[I]}{K_{I_2}}\right)} \tag{124}$$

or

$$\frac{E_L}{E_{Lmax}} = \frac{1}{\frac{K_{D_1}}{[L]}\left(1 + \frac{[I]}{K_{I_1}}\right) + \left(1 + \frac{[I]}{K_{I_2}}\right)} \tag{125}$$

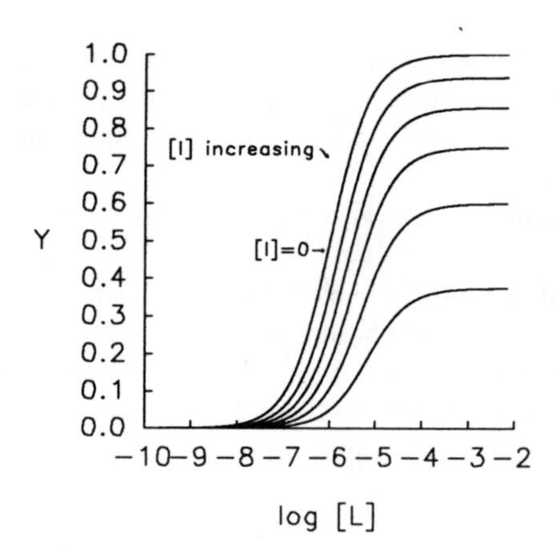

FIGURE 46

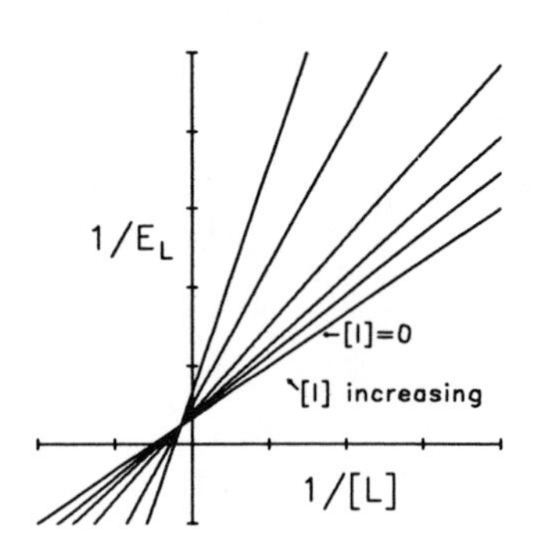

FIGURE 47

The double-reciprocal form of Equation 125 is

$$\frac{1}{E_L} = \frac{1}{[L]} \frac{K_{D_1}}{E_{Lmax}} \left(1 + \frac{[I]}{K_{I_1}}\right) + \frac{1}{E_{Lmax}} \left(1 + \frac{[I]}{K_{I_2}}\right) \qquad (126)$$

When the binding of I decreases the affinity for L and when the binding of L decreases the affinity for I (or when $K_{I1} < K_{I2}$, and $K_{D1} < K_{D2}$) Equations 125 and 126 produce the graphs shown in Figures 46 and 47. In Figure 46 a line drawn through the midpoints of the sigmoid curves slants upward to the left. In Figure 47 the double-reciprocal lines intersect above the 1/[L] axis and left of the $1/E_L$ axis.

When the binding of I increases the affinity for L and when the binding of L increases the affinity for I (or when $K_{I1} > K_{I2}$, and $K_{D1} > K_{D2}$) Equations 125 and 126 produce the graphs shown in Figures 48 and 49. In Figure 48

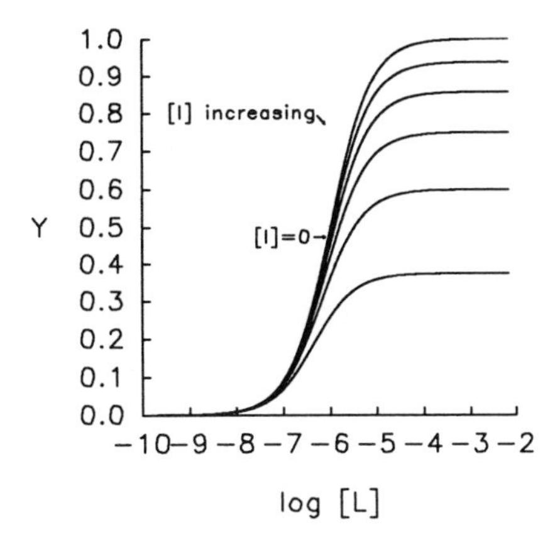

FIGURE 48

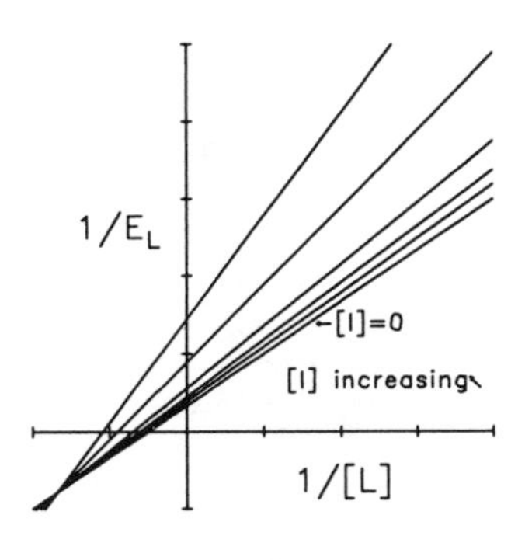

FIGURE 49

a line drawn through the midpoints of the sigmoid curves slants upward to the right and the lines intersect below the 1/[L] axis in Figure 49.

Heterotropic-cooperative noncompetitive inhibition will be detected in simple binding measurements. However, since binding detects both RL and RLI and is incapable of distinguishing between them the results can be confusing and misleading.

When the binding of a heterotropic-cooperative noncompetitive antagonist to its binding site reduces the affinity of the agonist binding site by at least 1000-fold, the binding of agonist is nearly eliminated when the antagonist binds. Similarly, when the binding of agonist to its binding site reduces the affinity of the antagonist binding site by at least 1000-fold, binding of the antagonist is nearly eliminated when the agonist binds. This relationship is described by the diagram in Figure 50. Here we have the extreme case of heterotropic-cooperative noncompetitive inhibition. When $K_{I_1} << K_{I_2}$ Equations 125 and 126 produce the graphs shown in Figures 51 and 52.

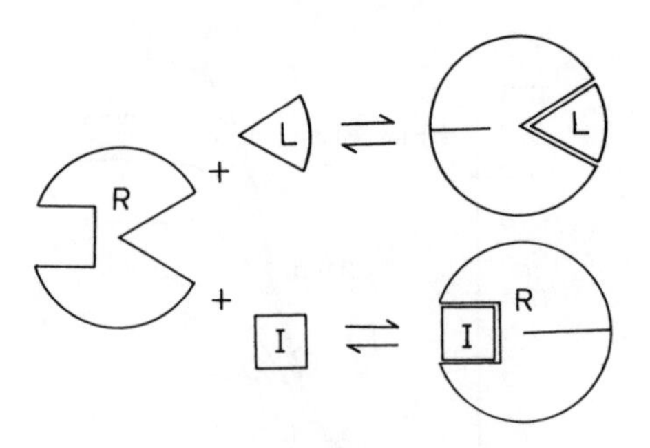

FIGURE 50

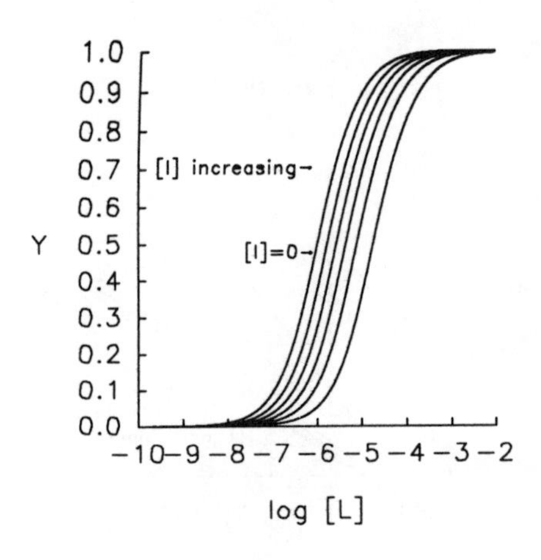

FIGURE 51

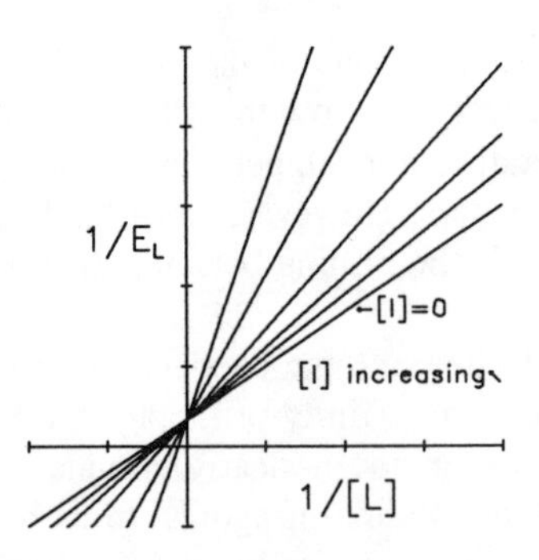

FIGURE 52

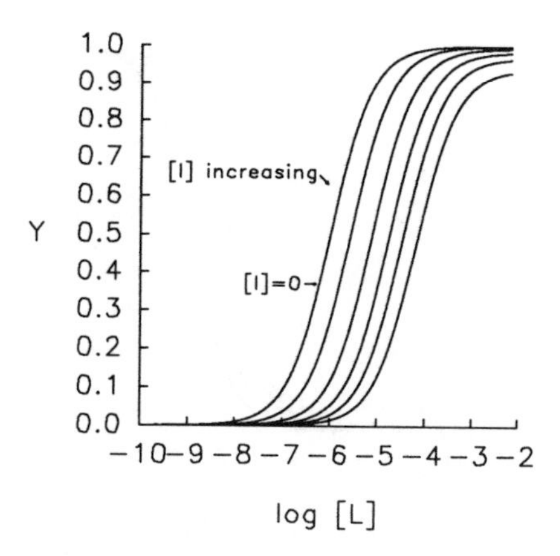

FIGURE 53

The extreme case of heterotropic-cooperative noncompetitive inhibition is indistinguishable from simple competitive inhibition except at very high inhibitor concentrations. This is shown in Figure 53. The difference in E_{Lmax} that occurs at very high inhibitor concentrations can easily be masked by experimental error. Even though this form of inhibition appears to be the same as simple competitive inhibition, except at very high inhibitor concentrations, its mechanism is completely different.

5. Allosteric Competitive Inhibition

Heterotropic cooperativity produces a third form of noncompetitive inhibition known as allosteric competitive inhibition. When the binding of the antagonist in Mechanism 12 produces a decrease in the affinity of the receptor for the agonist, while having no effect on the ability of the agonist to stimulate the receptor and no effect on the ability of the stimulated receptor to produce an effect, we have allosteric competitive inhibition. In this circumstance

$$\frac{E_L}{E_{Lmax}} = \frac{[RL] + [RLI]}{[R] + [RL] + [RI] + [RLI]} \tag{127}$$

Substituting from Equations 119 to 122 into Equation 127, and rearranging, gives

$$Y = \frac{1}{\dfrac{K_{D_1}}{[L]} \dfrac{\left(1 + \dfrac{[I]}{K_{I_1}}\right)}{\left(1 + \dfrac{[I]}{K_{I_2}}\right)} + 1} \tag{128}$$

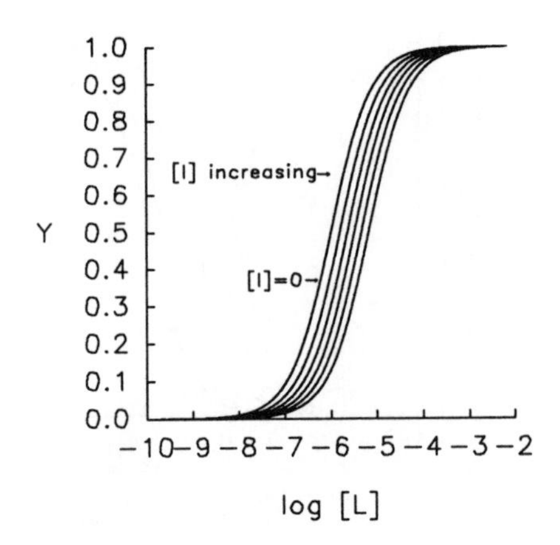

FIGURE 54

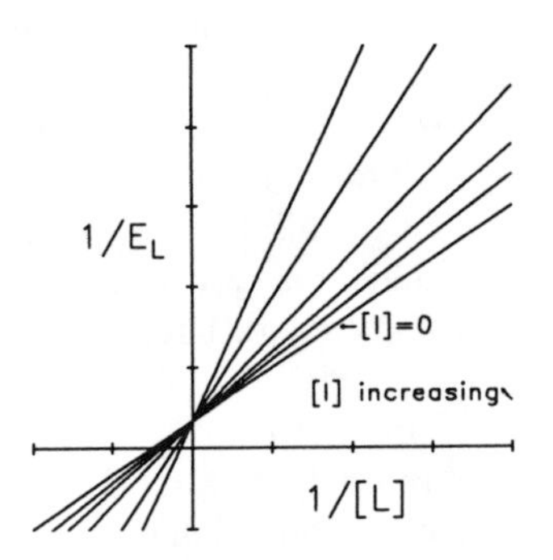

FIGURE 55

or the double-reciprocal form

$$\frac{1}{E_L} = \frac{1}{[L]} \frac{K_{D_1}}{E_{Lmax}} \frac{\left(1 + \frac{[I]}{K_{I_1}}\right)}{\left(1 + \frac{[I]}{K_{I_2}}\right)} + \frac{1}{E_{Lmax}} \tag{129}$$

For allosteric competitive inhibition to occur it is required that $K_{I_1} < K_{I_2}$. When this condition is met Equations 128 and 129 produce the graphs shown in Figures 54 and 55. Equations 128 and 129 differ from Equations 94 and 95 and from the extreme case of Equations 125 and 126 by yielding curved Schild and curved slope replots. This is shown in Figures 56 and 57.

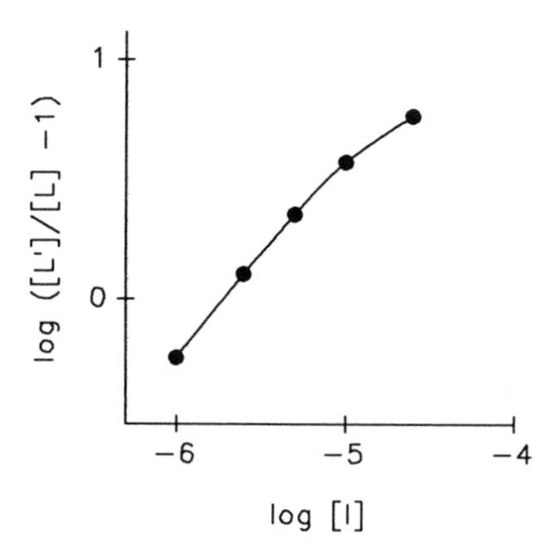

FIGURE 56

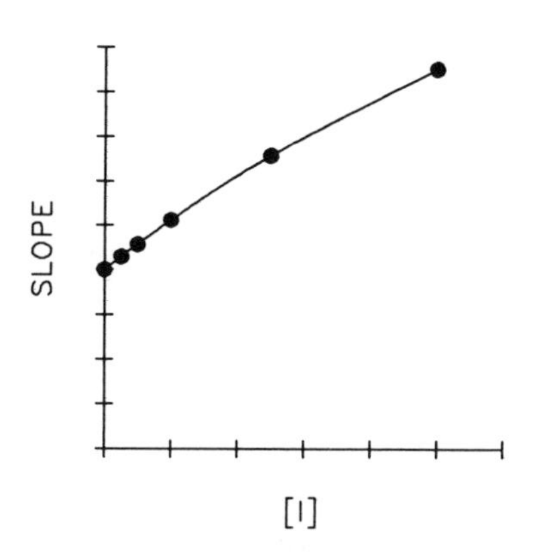

FIGURE 57

6. Allosteric Competitive Stimulation

When $K_{I_1} > K_{I_2}$ for Equations 128 and 129 the inhibitor is no longer an inhibitor. It is an allosteric competitive activator or stimulator. Therefore, we must change our designations from I for inhibitor to S for stimulator and Equations 128 and 129 yield the graphs shown in Figures 58 and 59.

7. Noncompetitive Stimulation

In addition to noncompetitive inhibition it is equally possible to have noncompetitive stimulation by all of the same mechanisms as already described for inhibition. With stimulation, however, the RL complex is not

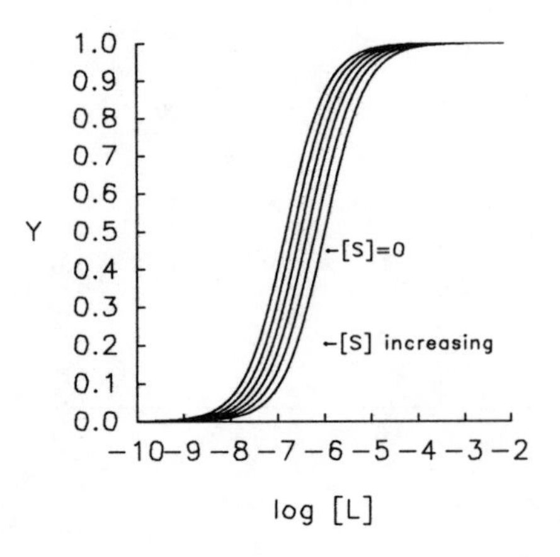

FIGURE 58

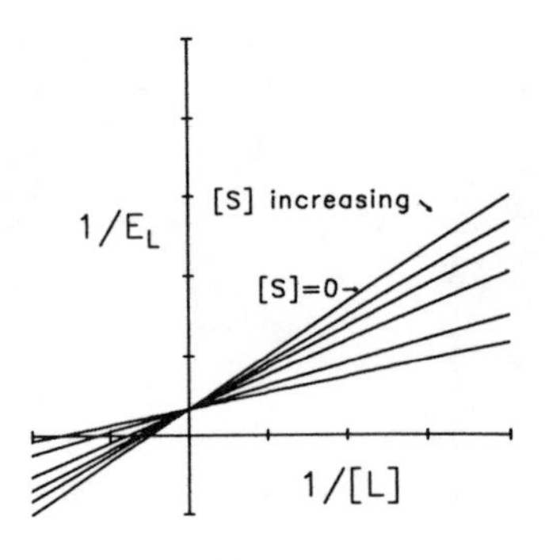

FIGURE 59

capable of producing receptor stimulation while the RLS complex is. This is shown in the modified version of Mechanism 12:

$$\begin{array}{ccc} R + L & \rightleftharpoons & RL \\ + & & + \\ S & & S \\ \uparrow\downarrow & & \uparrow\downarrow \\ RS + L & \rightleftharpoons & RLS \end{array} \qquad \text{(Mechanism 13)}$$

where RLS is the stimulated form of the receptor. Here

$$\frac{E_L}{E_{Lmax}} = \frac{[RLS]}{[R] + [RL] + [RS] + [RLS]} \tag{130}$$

$$K_{D_1} = \frac{[R]\,[L]}{[RL]} \tag{131}$$

$$K_{D_2} = \frac{[RS]\,[L]}{[RLS]} \tag{132}$$

$$K_{S_1} = \frac{[R]\,[S]}{[RS]} \tag{133}$$

and

$$K_{S_2} = \frac{[RL]\,[S]}{[RLS]} \tag{134}$$

Substitution and rearrangement yields

$$Y = \frac{1}{\frac{K_{D_2}}{[L]}\left(1 + \frac{K_{S_1}}{[S]}\right) + \left(1 + \frac{K_{S_2}}{[S]}\right)} \tag{135}$$

or the double-reciprocal form

$$\frac{1}{E_L} = \frac{1}{[L]}\frac{K_{D_2}}{E_{Lmax}}\left(1 + \frac{K_{S_1}}{[S]}\right) + \frac{1}{E_{Lmax}}\left(1 + \frac{K_{S_2}}{[S]}\right) \tag{136}$$

When $K_{S_1} = K_{S_2}$ we have simple noncompetitive stimulation and Equations 135 and 136 plot as shown in Figures 60 and 61. When $K_{S_1} > K_{S_2}$, Equations 135 and 136 plot as shown in Figures 62 and 63. When $K_{S_1} < K_{S_2}$, Equations 135 and 136 plot as shown in Figures 64 and 65. Then, similar to the situation where $K_{I_1} << K_{I_2}$, when $K_{S_1} << K_{S_2}$ we have the extreme case of heterotropic-cooperative noncompetitive stimulation which graphs as shown in Figure 66.

Noncompetitive stimulation presents a dilemma from the standpoint that it will be difficult, if not impossible, to determine which ligand is the stimulator and which is the agonist under many circumstances. It is equally correct in most circumstances to refer to both L and S as agonists in a mechanism where receptor stimulation requires the binding of two different agonists to the same receptor.

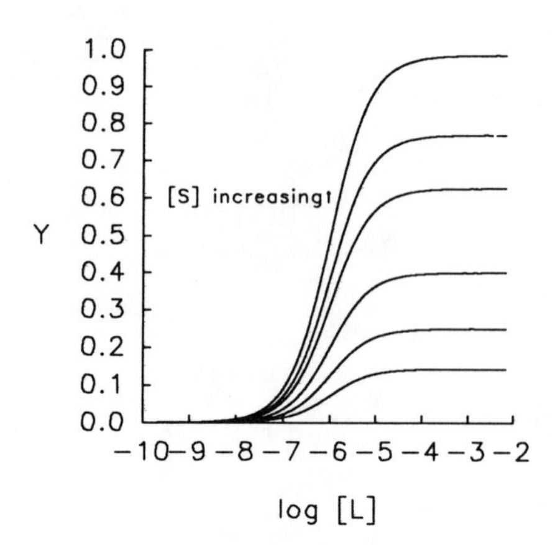

FIGURE 60

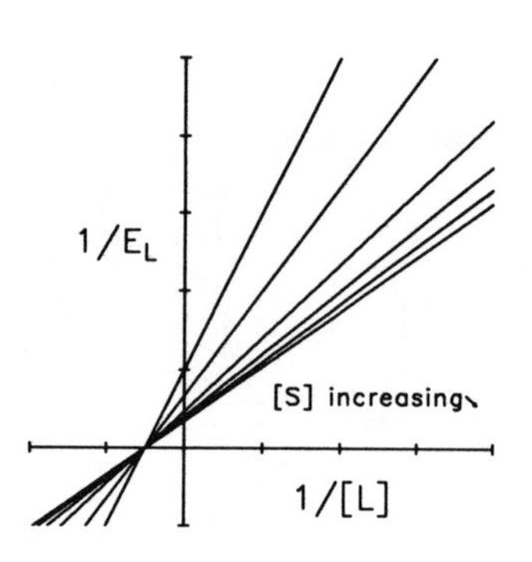

FIGURE 61

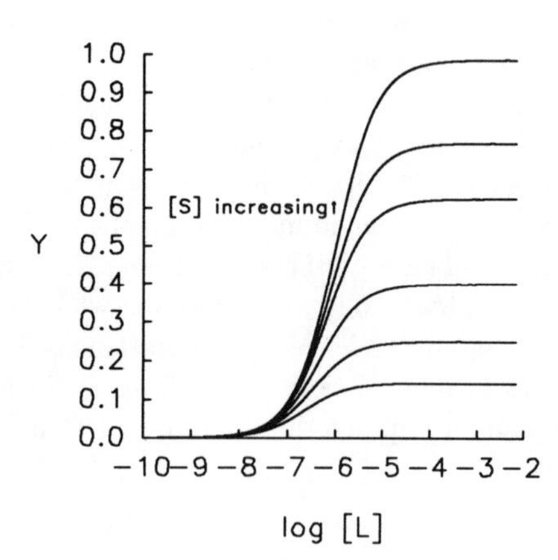

FIGURE 62

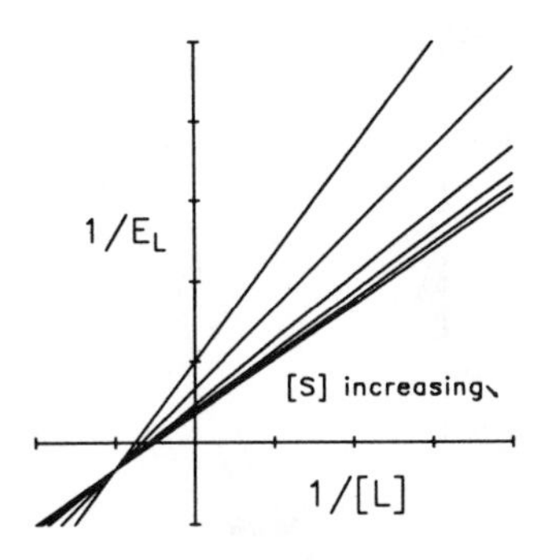

FIGURE 63

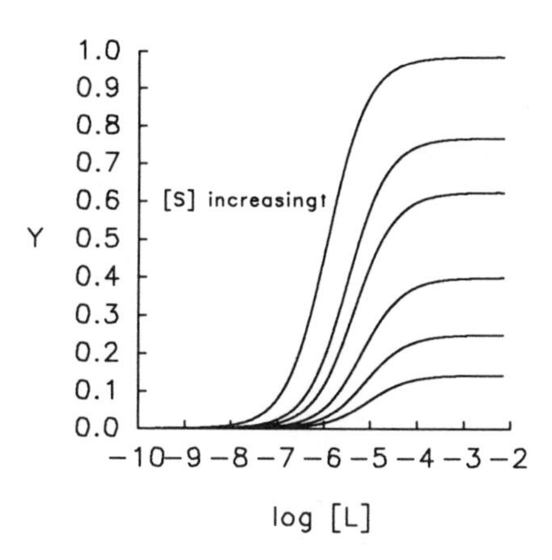

FIGURE 64

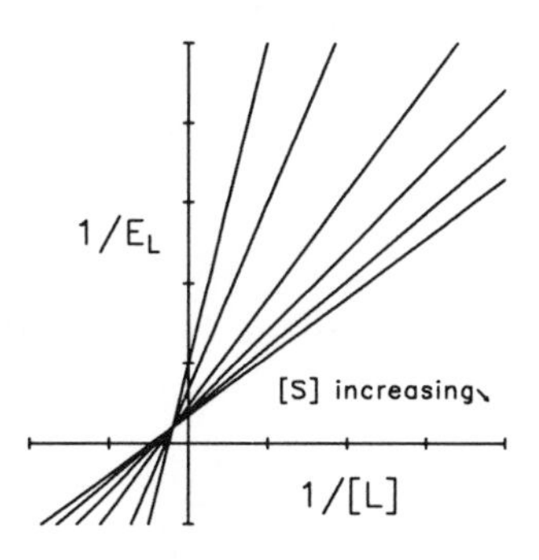

FIGURE 65

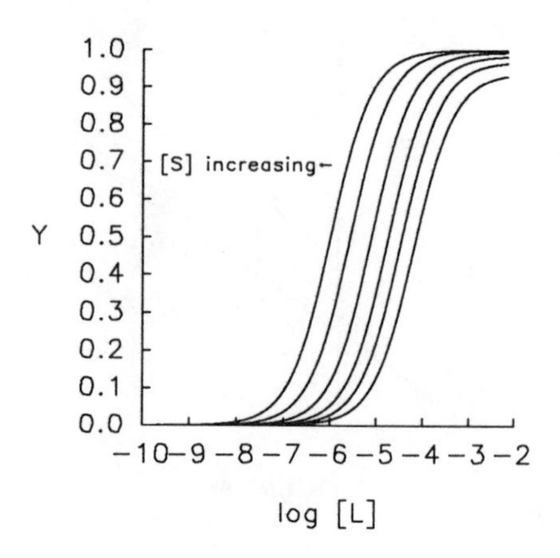

FIGURE 66

8. Homotropic Noncompetitive Inhibition

The noncompetitive inhibitor does not have to be a different molecule than the agonist. It is quite possible that a single molecule can function as both an agonist and an antagonist on the same receptor. This requires that the receptor have two binding sites for the ligand. Binding to one of the sites results in receptor stimulation. Binding to the other site results in a blockade of receptor stimulation or a blockade of the effect of receptor stimulation. This is illustrated by Mechanism 14:

$$\begin{array}{ccc} R + L & \rightleftharpoons & RL_S \\ + & & + \\ L & & L \\ \uparrow\downarrow & & \uparrow\downarrow \\ RL_I & \rightleftharpoons & RLL_I \end{array} \qquad \text{(Mechanism 14)}$$

where the subscripts S and I indicate stimulated and inhibited complexes, respectively. Here

$$\frac{E_L}{E_{Lmax}} = \frac{[RL_S]}{[R] + [RL_S] + [RL_I] + [RLL_I]} \tag{137}$$

$$K_{D_1} = \frac{[R]\,[L]}{[RL_S]} \tag{138}$$

$$K_{D_2} = \frac{[RL_I]\,[L]}{[RLL_I]} \tag{139}$$

$$K_{I_1} = \frac{[R]\,[L]}{[RL_I]} \tag{140}$$

and

$$K_{I_2} = \frac{[RL_S]\,[L]}{[RLL_I]} \tag{141}$$

Substitution and rearrangement yield

$$Y = \frac{1}{\frac{K_{D_1}}{[L]}\left(1 + \frac{[L]}{K_{I_1}}\right) + \left(1 + \frac{[L]}{K_{I_2}}\right)} \tag{142}$$

or the double-reciprocal form

$$\frac{1}{E_L} = \frac{1}{[L]}\frac{K_{D_1}}{E_{Lmax}}\left(1 + \frac{[L]}{K_{I_1}}\right) + \frac{1}{E_{Lmax}}\left(1 + \frac{[L]}{K_{I_2}}\right) \tag{143}$$

When all the affinity constants are equal to one another Equations 142 and 143 give the graphs shown in Figures 67 and 68. Generally, the inhibition affinities will be equal to or larger than the agonist affinities for mechanisms of homotropic noncompetitive inhibition. Homotropic-allosteric competitive inhibition mechanisms are also possible.

9. Uncompetitive Antagonism

Uncompetitive antagonists can only combine with the RL complex to produce an RLI complex that is incapable of leading to the production of a physiological effect. This form of antagonism can be represented by Mechanism 15 and Figure 69.

$$\begin{array}{c} R + L \rightleftharpoons RL \\ \quad\quad\quad + \\ \quad\quad\quad I \\ \quad\quad\quad \uparrow\downarrow \\ \quad\quad\quad RLI \end{array} \qquad \text{(Mechanism 15)}$$

The equations that describe uncompetitive inhibition are

$$\frac{E_L}{E_{Lmax}} = \frac{1}{\frac{K_D}{[L]} + \left(1 + \frac{[I]}{K_I}\right)} \tag{144}$$

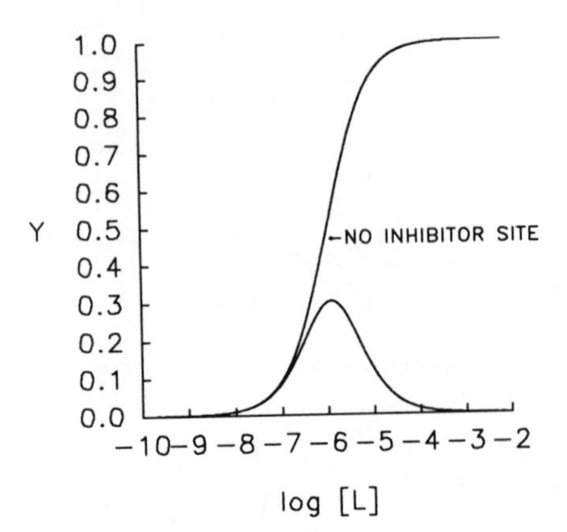

FIGURE 67

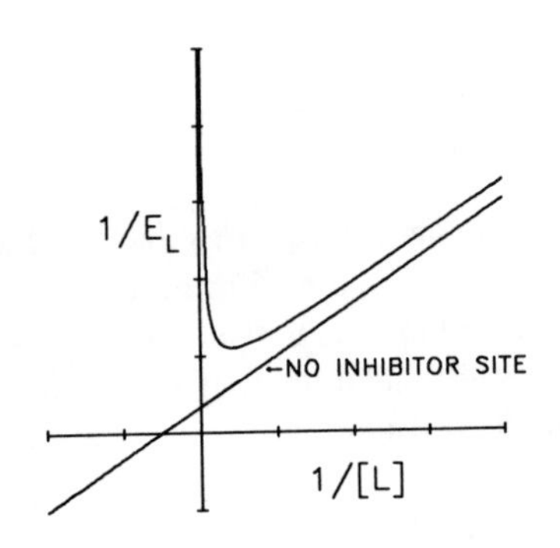

FIGURE 68

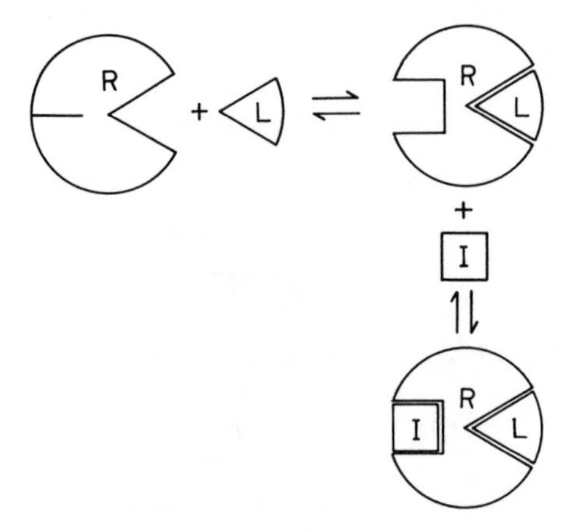

FIGURE 69

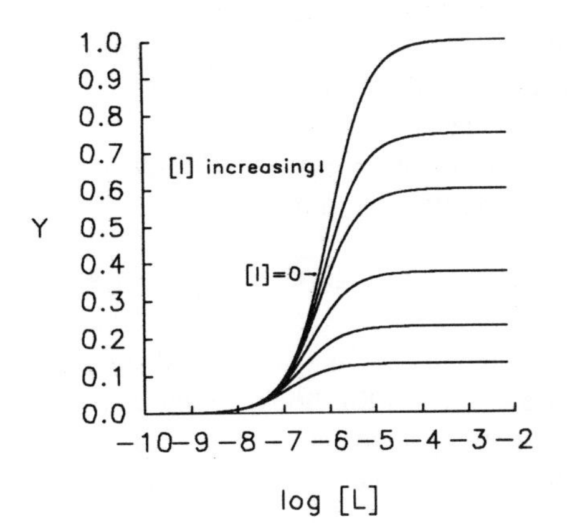

FIGURE 70

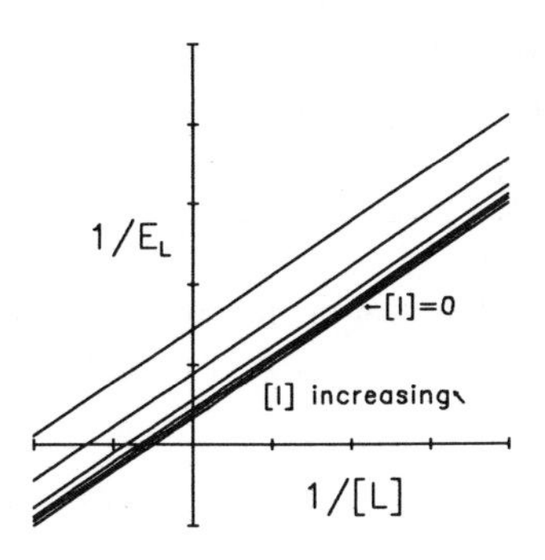

FIGURE 71

or the double-reciprocal form

$$\frac{1}{E_L} = \frac{1}{[L]}\frac{K_D}{E_{Lmax}} + \frac{1}{E_{Lmax}}\left(1 + \frac{[I]}{K_I}\right) \tag{145}$$

Plotting Equation 144 yields Figure 70: where the maximum extent of the sigmoid curve decreases as the inhibitor concentration increases. A line drawn through the abscissal midpoints of the sigmoid curves slants upward to the right. Plotting Equation 145 yields Figure 71. Here a series of different inhibitor concentrations produces a series of parallel lines in which the inhibitor causes increases in both intercepts without affecting the slope. Again, since simple ligand binding measurements are incapable of detecting the difference between RL and RLI, uncompetitive antagonists will not be detected. If we have a circumstance where two different agonists are required, as described in the previous section, and if agonist 2 can only bind after agonist 1 has bound to produce a stimulated RL_1L_2 complex, then we have a mechanism of uncompetitive stimulation. This is the same as the situation described for Mechanism 8 and Equation 76 in Chapter 7, with the exception that the two ligands are different molecules.

III. CONCLUSION

To conclude this chapter on antagonism and stimulation, and to conclude the receptor kinetics portion of this book, we have shown several possibilities for aberrant or complex behavior in receptor kinetics. All the graphs shown are theoretical graphs where there is no experimental error involved in the numbers. Real-life receptor data always has experimental error, which cannot be eliminated. Experimental error can very often obscure such things as the location of intersects, the slopes of lines, and whether a line is straight or curved. The limitations imposed on the researcher by experimental error make an in-depth understanding of the mechanisms for producing complex receptor kinetics very important. Several different analyses should always be done, even on systems that appear to give simple kinetics. Then, when unusual behavior is observed, the researcher has two options to sort out what is going on. One can assume elaborate kinetic models to describe the behavior and derive equations by which to test the models, or one can simplify the system in order to simplify the kinetics. Of these two options the latter is the preferred option when feasible.

Part II: Enzymes

Chapter 9

INTRODUCTION TO ENZYMES

I. ENZYME CATALYSIS AND SPECIFICITY

Enzymes are biological macromolecules, generally proteins, (although some RNA enzymes are known) that increase the rates of (catalyze) selected chemical reactions. There is a separate enzyme, or sometimes more than one enzyme, to catalyze nearly every chemical reaction that occurs in, or is associated with, a living organism. Enzymes have the added property of providing for selectivity of reactants and products of the chemical reaction they catalyze.

Consider a molecule such as D-glucose (Figure 72). Suppose that we want to convert the 3-position of D-glucose to an acetate ester by reaction with acetic acid. Combining glucose and acetic acid alone is probably insufficient to produce a reaction. We need to provide the necessary conditions to cause the reaction to proceed, i.e., dilute mineral acid. After some time has been allowed for the reaction to proceed, we find that not only did we produce 3-acetyl-D-glucose as a product, we also produced acetyl derivatives of each of the other hydroxyl groups of the molecule as well as di-, tri-, tetra-, and penta-acetate derivatives. Clearly, the dilute mineral acid catalyst was not capable of providing specificity of products. We could have adjusted conditions to limit the number of unwanted products. However, we would still find it necessary to purify, out of the reaction mixture, the one product that we wanted.

Now suppose that we mix some L-glucose, D-galactose, and some D-ribose with the D-glucose and run the same reaction. Again all we want is 3-acetyl-D-glucose as our product. What we will get is a mixture of all possible acetylation products of all four sugars. Now instead, let us mix acetic acid, D-glucose, L-glucose, D-ribose, and D-galactose at pH 7 and add the imaginary enzyme D-glucose-3-acetylase. After allowing time for the reaction to come to completion we will find 3-acetyl-D-glucose, L-glucose, D-ribose and D-galactose in the mixture, along with some unreacted acetic acid. The enzyme has provided for both reactant (called the substrate in enzyme terminology) and product specificity. Another thing that the enzyme accomplished was that it allowed the reaction to proceed at neutral pH as compared with low pH for the nonenzymatic reaction.

How can enzymes accomplish increases in rates of chemical reactions and provide for both product and substrate specificity, all at neutral pH and body temperature? They do so by providing a binding site or pocket on their surface into which the substrate molecule fits with precise shape and chemical constraints. This is the same as the binding interaction between receptors and

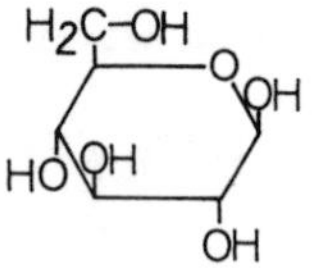

FIGURE 72

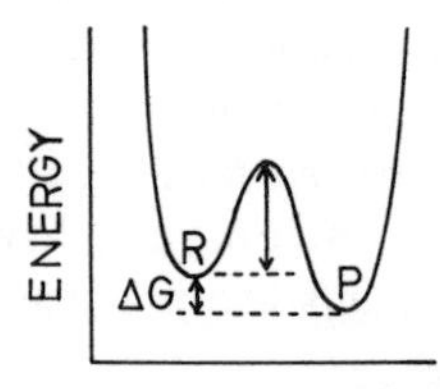

FIGURE 73

ligands. Enzymes, however, go one step further than receptors and provide within that binding site a micro-environment that represents the optimal condition necessary for the desired chemical reaction to occur.

To follow the example of D-glucose-3-acetylase: of the four sugars in our mixture only D-glucose will fit properly into the binding site (active site) of the enzyme. Once bound, the hydroxyl group on the 3-position, and only the hydroxyl group on the 3-position, is located in a micro-environment that simulates the dilute mineral acid. The enzyme also binds a molecule of acetic acid (as a second substrate) such that it is located and oriented just right to react with the 3-hydroxyl of the D-glucose to produce 3-acetyl-D-glucose.

II. REACTION RATES

Enzymes cannot catalyze reactions that are impossible without the enzyme. They can, however, provide for specificity of substrates and products that cannot be easily achieved in any other way. By providing favorable micro-environments, and proper orientation of substrates in these micro-environments, enzymes accomplish a lowering of the activation energy of the reaction. This results in increased reaction rates. Reaction activation energies are the amount of energy that must be added to the system in order to get the reaction to proceed. These can most effectively be illustrated by the energy diagram in Figure 73. In this diagram both the reactants and products are in their ground energy states. In order for reactants to become products sufficient energy must be added to the system to convert the reactants to the transitional energy state. This is the point at the top of the energy hill (or energy barrier) that separates the reactants from the products. Once this transition energy state has been achieved the system can lose energy by going back downhill in either direction to become products or to remain reactants.

It is not necessary for every molecule in the mixture to obtain activation energy for the reaction to proceed. Molecules are constantly vibrating and bumping into one another. Collisions between molecules often involve energy transfers such that one molecule will lose energy to the other when two molecules collide. This provides a means for individual molecules to accumulate sufficient energy to reach the transition state and undergo reaction. If we warm the reaction mixture we increase the amount of energy available. Individual molecules are then able to obtain sufficient activation energy and undergo reaction with increased frequency. **When we warm the reaction mixture we increase the rate of the chemical reaction.**

By reducing the height of the activation energy barrier enzymes reduce the amount of energy necessary for a molecule to reach the transition state. This allows the reaction to proceed faster at lower temperatures. In this way, enzymes are capable of very large increases in chemical reaction rates. An energy diagram can be compared to a road over a mountain with cities on either side connected by the road. Commodities traveling from one city to the other must be hauled over the mountain by truck. If you have ever followed a loaded truck up a steep mountain grade you have an idea of how difficult obtaining the necessary activation energy can be. Now suppose that the two cities pool their resources and dig a tunnel through the mountain. Trucks are no longer required to climb the steep grades on the way up and wear out their brakes on the way down. Thus, transit between the two cities is quicker and more efficient. The enzyme acts like the tunnel. The source and destination are still the same, but the barrier separating them has changed.

III. ANALYSIS OF ENZYME-CATALYZED REACTIONS

There are many ways for measuring enzyme reaction rates. The most favorable situations are those where the product is a different color than the substrate or the product has a different absorption spectrum than the substrate. When such differences occur, the reaction can be performed in a cuvette in a spectrophotometer and the change in absorption properties can be followed as the reaction proceeds. Changes in absorption can be directly related to changes in substrate and product concentrations. Similarly, if the substrate or product is fluorescent, or if the substrate fluorescence has a different spectrum than the product fluorescence, the change in fluorescence can be followed as the reaction proceeds. Fluorescence changes can be related to changes in substrate and product concentrations. Some enzyme-catalyzed reactions produce light as one product of the reaction (bioluminescence). With these reactions the intensity of light emission is a direct measure of the reaction rate. Some enzyme-catalyzed reactions result in the production of a gaseous product such as CO_2. CO_2 can be driven out of solution by acidification and trapped in a separate alkaline solution. If the carbon in the CO_2 is ^{14}C it can then be quantified by scintillation counting. Sometimes the product of an enzyme-catalyzed reaction is very large or very small compared with the substrate. When this occurs, techniques such as differential centrifugation and ultrafiltration, in combination with radiolabeling or other forms of labeling, can be employed for analysis.

When no useful color, fluorescence, or size changes, etc., are available it is often possible to couple the reaction to other enzymes that will produce such changes. In coupled reactions the enzyme or substrate of interest is limiting in the reaction mixture. All the coupling reagents and enzymes are at saturating levels. An example of a coupled reaction is the enzyme hexokinase, which converts glucose plus ATP to glucose-6-phosphate plus ADP. After some reaction time has elapsed, a firefly luciferin-luciferase mixture is added to the hexokinase reaction mixture. The ATP that remains in the mixture reacts with the luciferin-luciferase to produce a flash of light. The intensity of the light flash is proportional to the amount of ATP that was present.

Sometimes it is possible to produce colored products by using reagents that react specifically with the substrate or product of a reaction. These would be used much like the luciferin-luciferase reaction. When all other possibilities are impractical, products or substrates may be isolated from reaction mixtures by chromatographic techniques and quantified by appropriate methods.

The equations that we will derive for the analysis of enzyme-catalyzed reactions are essentially the same equations as those we have derived for receptor-ligand interactions. In many instances they can be used in the same ways. Enzymes cannot, under most circumstances, be directly measured by their abilities to bind their substrates as we do with receptor-ligand interactions. This is because as soon as the substrate binds with the enzyme it is converted to product. It is much more straightforward to measure enzymes by following the rate of loss of substrate or the rate of formation of product. Therefore, instead of relating the amount of bound ligand to the concentration of unbound ligand, we have to relate the rate of reaction to the concentration of substrate. Only in situations where more than one substrate, or a substrate and a cofactor, are required for the reaction and one of these is omitted, can ligand binding analysis be employed to study enzyme-substrate interactions. In the next two chapters we will address the differences between the analysis of enzyme-substrate interactions and receptor-ligand interactions and derive equations that describe enzyme kinetics.

Chapter 10

SIMPLE ENZYME SUBSTRATE INTERACTIONS

I. NONCATALYZED REACTION KINETICS

Consider a chemical reaction that is not enzyme-catalyzed, such as

$$R \rightleftharpoons P \qquad \text{(Mechanism 16)}$$

Here R is the reactant and P is the product. When we start the reaction we have 100% R and 0% P. If we plot the concentrations of R and P over time we generate the graph shown in Figure 74. We can see, from Figure 74, that as time goes by [R] decreases and [P] increases. Eventually, there is no further change in [R] and no further change in [P] with time. This does not mean that the reaction has stopped, far from it in fact. The double arrow in Mechanism 13 indicates that the reaction is reversible. As soon as there is any P in the reaction mixture it can undergo conversion back to R. The higher the concentration of P the higher the number of molecules of P being converted back to R in any period of time. Similarly, the lower the concentration of R the lower the number of molecules of R being converted to P in any period of time. Eventually the reaction reaches a condition where the number of P being converted to R in any period of time equals the number of R being converted to P. Once this condition has been reached there is no further net change in the reaction mixture. The reaction has reached equilibrium.

The relative concentrations of R and P at equilibrium are determined by the relative energy contents of the two molecules. If R and P are equal in energy content they will be equal in concentration at equilibrium. If P is lower than R in energy content, then P will be higher in concentration than R at equilibrium. The greater the difference in energy content between R and P the greater the difference in their concentrations at equilibrium. Remembering that the concentration ratio at equilibrium is equal to the equilibrium constant (K), the relationship between energy difference and concentrations at equilibrium has already been described in Equation 10.

$$\Delta G = \Delta G^{\circ} + RT\ln K_D \qquad (10)$$

Returning to the analogy of the two cities on opposite sides of a mountain connected by a tunnel (my *Tale of Two Cities* analogy), when the two cities are at the same elevation, trucks traveling from one city to the other have a level journey. If one city is at a lower elevation than the other, trucks traveling to the lower city can coast all the way while they need to burn fuel on the return, uphill, trip. The burning of fuel is a good example of an uncatalyzed

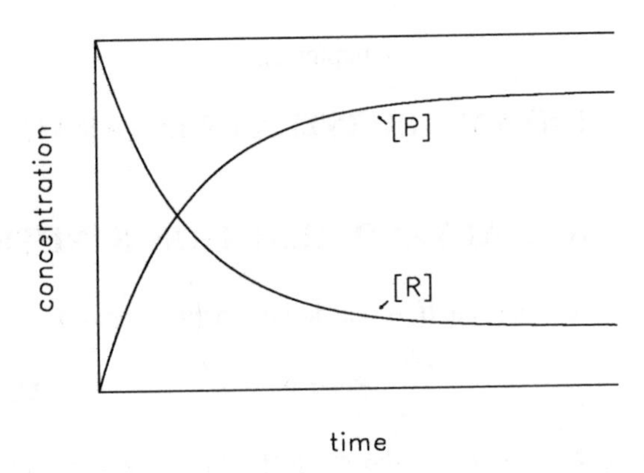

FIGURE 74

chemical reaction where the reactants and products have very different energy contents. The fuel is a hydrocarbon having the general chemical composition of $C_nH_{(2n+2)}$. This reacts with molecular oxygen (O_2) from the atmosphere, producing CO_2 and H_2O. The reaction is highly exothermic, meaning that it is associated with a large decrease in energy content. Therefore, the equilibrium is very much in favor of the products CO_2 and H_2O. The fuel is burned in the engine where there is a mechanism for converting the released chemical energy into kinetic energy to make the truck go. The fuel is stored in the fuel tank of the truck where there is also oxygen. What prevents the reaction from occurring in the fuel tank instead of in the engine? The answer is that we do not provide the necessary energy of activation to start the reaction when the fuel is in the tank. In the engine combustion chamber the oxygen-fuel mixture is brought into contact with an ignition source, which provides the necessary energy of activation to start the reaction. Then, since the reaction is so highly exothermic, it provides its own source of activation energy to sustain the reaction.

If we look at Figure 74 we see that the slope of the [R] vs. time curve becomes shallower as the reaction time increases. Similarly, the slope of the [P] vs. time curve becomes shallower as the reaction time increases. If we draw a horizontal line through the point where the two curves cross we see that the curves are mirror images of one another in relation to this line. The slopes of the curves represent the changes in concentration divided by the changes in time. For R the slope is -d[R]/dt, or the negative change in [R] divided by the change in time. For P the slope is d[P]/dt, or the change in [P] divided by the change in time. Since the curves are symmetrical to one another we can see that for any period of time (dt)

$$\frac{-d\,[R]}{dt} = \frac{d\,[P]}{dt} \tag{146}$$

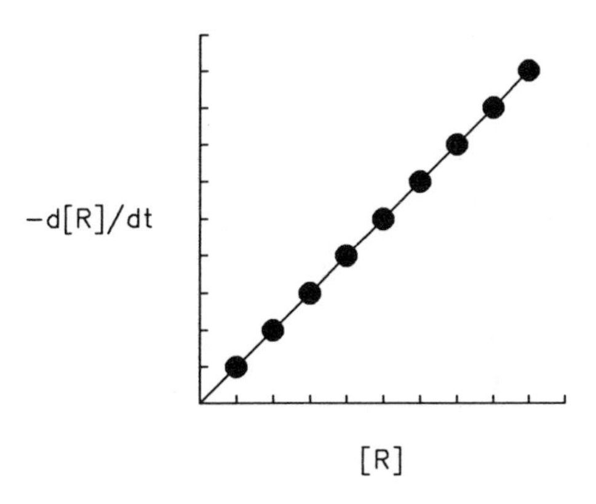

FIGURE 75

where -d[R]/dt and d[P]/dt are the rates of the chemical reactions. Reaction rates are composed of a concentration component and a time component. At time zero, when the concentration of R is high, the reaction rate is high. As the reaction time increases the concentration of R decreases and, correspondingly, the reaction rate decreases. This discussion suggests that the concentration of reactant is proportional to the reaction rate. This is indeed correct. The proportionality constant that relates the reactant concentration to the reaction rate is the rate constant (k). Mechanism 16 can now be redisplayed with rate constants added

$$R \underset{k_{-1}}{\overset{k_{+1}}{\rightleftharpoons}} P \qquad \text{(Mechanism 17)}$$

Here k_{+1} is the rate constant for the reaction R going to P and k_{-1} is the rate constant for the reaction P going to R. This gives rise to the relationship

$$\frac{-d\,[R]}{dt} = k_{+1}\,[R] \tag{147}$$

Now, if we plot -d[R]/dt vs. [R] at time zero (i.e., the initial rate of the reaction vs. the initial concentration of the reactant) for a series of reactions in which the initial reactant concentration is varied, we get the relationship shown in Figure 75. Figure 75 illustrates what we already know from Equation 147: -d[R]/dt is related to [R] by a linear relationship where the slope is k_{+1} and the intercept on the -d[R]/dt axis is zero. Restating what we see in Figure 75, the larger the initial concentration of R the faster the initial reaction rate.

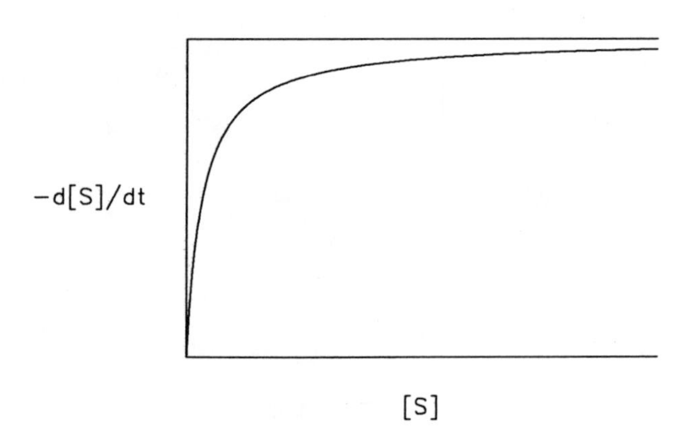

FIGURE 76

II. ENZYME-CATALYZED REACTION KINETICS

If we plot the reaction rate vs. [R] (or [S] for substrate concentration for an enzyme-catalyzed reaction we get the relationship shown in Figure 76. Here we see that -d[S]/dt and [S] are not linearly related. The difference between an enzyme-catalyzed reaction and a reaction that is not enzyme-catalyzed is that the reactant (or substrate) must form a complex (or bind) with the enzyme before the reaction can occur. Therefore, the actual reactant species is not S, but rather it is the enzyme-substrate complex (ES). In the enzyme-catalyzed reaction the amount of enzyme is constant and we reach a point as we increase [S] where [S] becomes so high that nearly all the enzyme is in the ES form all the time. Further increases in [S] cannot increase the rate of the reaction beyond this point since the enzyme is working to catalyze the reaction at its maximum capacity. The enzyme has reached saturation.

What we need is a mathematical expression that relates the substrate concentration to the rate of the reaction and, which describes the behavior seen in Figure 76. Unlike receptor binding measurements, the substrate concentration (which is analogous to the ligand or agonist concentration) will not be static in an enzyme-catalyzed reaction. It is continually decreasing as it is converted to product. For this reason, we need to choose a point in time when we know accurately the concentration of S. The only time when we do know [S] accurately is at the instant when we mix the enzyme and substrate together. This is time zero. If we measure the rate of the reaction as close to time zero as it is possible to do we have the initial rate. This is also known as the initial velocity (V) of the reaction.

We must first modify Mechanism 17 to reflect the role of the enzyme in the reaction

$$E + S \rightleftharpoons ES \rightleftharpoons EP \rightarrow E + P \qquad \text{(Mechanism 18)}$$

The enzyme and substrate combine to form the ES complex. Then, the substrate is converted to product in the enzyme active site. Finally, the enzyme-product complex dissociates. Each of the steps in this mechanism is reversible. We did not include a reverse arrow for the last step since all the measurements we will make with an enzyme-catalyzed reaction will be the initial rate where [P] = zero. We also can ignore the middle step, conversion of ES to EP. This step is kinetically indistinguishable in the mechanism. This is the rapid equilibrium assumption. There may be many steps involved in the conversion of ES to E + P, however, they all can be lumped together under a single rate constant. The slowest step in the sequence will determine the magnitude of this constant. We can then assign rate constants to the various steps to yield Mechanism 19:

$$E + S \underset{k_{-1}}{\overset{k_{+1}}{\rightleftharpoons}} ES \overset{k_{+2}}{\rightharpoonup} E + P \qquad \text{(Mechanism 19)}$$

Next, since ES is the true reactant in this mechanism, we know that

$$V = k_{+2}\,[ES] \qquad (148)$$

However, neither k_{+2} nor [ES] can be evaluated under most circumstances, so V must be expressed in terms that we can measure. We know that the rate of formation of ES or

$$\frac{d\,[ES]}{dt} = k_{+1}\,[E]\,[S] \qquad (149)$$

We also know that

$$[E_T] = [E] + [ES] \qquad (150)$$

Rearranging Equation 150 and substituting into Equation 149 we get

$$\frac{d[ES]}{dt} = k_{+1}\,([E_T] - [ES])\,[S] \qquad (151)$$

Next, the equation for the rate of destruction of ES is obtained by taking the sum of the reactions leading away from ES from Mechanism 19:

$$-\frac{d[ES]}{dt} = k_{-1}\,[ES] + k_{+2}\,[ES] \qquad (152)$$

Then we must establish the rule that enzyme-catalyzed reactions are always set up such that $[S_T] >> [E_T]$. Under this condition we can see, from Figure

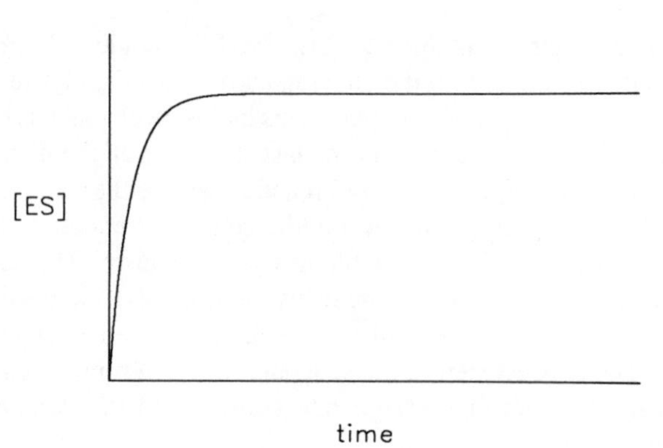

FIGURE 77

77, that a condition will rapidly be achieved where the concentration of ES doesn't change over time. In other words, both d[ES]/dt and -d[ES]/dt are equal to zero during the early phase of the reaction. This is the period in which we measure the initial rate. This is the steady state approximation. It allows us to combine Equations 151 and 152, yielding

$$k_{+1}\,([E_T] - [ES])\,[S] = k_{-1}\,[ES] + k_{+2}\,[ES] \tag{153}$$

Rearranging gives

$$\frac{k_{-1} + k_{+2}}{k_{+1}} = \frac{([E_T] - [ES])\,[S]}{[ES]} \tag{154}$$

Here we can define a new constant, the Michaelis constant (K_M)

$$K_M = \frac{k_{-1} + k_{+2}}{k_{+1}} \tag{155}$$

and

$$K_M = \frac{[E_T]\,[S] - [ES]\,[S]}{[ES]} \tag{156}$$

or

$$K_M = \frac{[E_T]\,[S]}{[ES]} - [S] \tag{157}$$

Keeping in mind that

$$[S_T] = [S] + [ES] \tag{158}$$

and since we have established the condition that $[S_T] >> [E_T]$ it follows that $[ES]$ will represent an insignificantly small proportion of $[S_T]$. Therefore, $[S_T]$ will equal $[S]$.

We can rearrange Equation 157 to solve for [ES]:

$$[ES] = \frac{[E_T]\,[S]}{K_M + [S]} \tag{159}$$

Recalling Equation 148

$$V = k_{+2}\,[ES] \tag{148}$$

it follows that the maximum rate (V_{max}) of the reaction occurs when all of E is in the form ES ($[ES] = [E_T]$), or

$$V_{max} = k_{+2}\,[E_T] \tag{160}$$

Next, we can multiply both sides of Equation 159 by k_{+2} to obtain

$$k_{+2}\,[ES] = \frac{k_{+2}\,[E_T]\,[S]}{K_M + [S]} \tag{161}$$

and substituting from Equations 148 and 160 we have

$$V = \frac{V_{max}\,[S]}{K_M + [S]} \tag{162}$$

the Michaelis-Menten equation for enzyme kinetics.[2]

The Michaelis-Menten equation has the same form as the Langmuir binding isotherm (Equation 18), which we derived in Chapter 2, where V replaces B_L, [S] replaces [L], and K_M replaces K_D. **The Michaelis-Menten equation can be used in many of the same ways as the Langmuir binding isotherm as long as one keeps in mind that there were several assumptions and constraints employed in its derivation which must hold true.** If these conditions are not met the Michaelis-Menten equation becomes invalid.

What does the Michaelis constant tell us about the enzyme-substrate interaction? K_M has the units of concentration, just like the K_D. If we arbitrarily set [S] equal to K_M, then Equation 162 becomes

$$V = \frac{V_{max}}{2} \tag{163}$$

The K_M is equal to the substrate concentration that gives one half of the maximum rate. The K_M is related to the dissociation constant (K_D) for the enzyme-substrate interaction as follows. From Mechanism 19

$$E + S \underset{k_{-1}}{\overset{k_{+1}}{\rightleftharpoons}} ES \overset{k_{+2}}{\rightharpoonup} E + P \qquad \text{(Mechanism 19)}$$

we know that

$$K_D = \frac{k_{-1}}{k_{+1}} = \frac{[E]\,[S]}{[ES]} \tag{164}$$

After rearranging Equation 155 and substituting from Equation 164 we have

$$K_M = K_D + \frac{k_{+2}}{k_{+1}} \tag{165}$$

The relationship between K_M and K_D is dependent upon the relationship between k_{+2} and k_{+1}. If $k_{+2} << k_{+1}$, then K_M will be equal to K_D. When $k_{+2} << k_{+1}$ the rate of formation of the ES complex from E and S is very fast compared with the rate of formation of E and P from ES. This is normally true. However, we can rarely be certain that $k_{+2} << k_{+1}$ and, when it is not, then K_M will not be equal to K_D.

All the linear transformations that we derived in Chapter 2 for equilibrium binding are valid for use with enzyme kinetic data. Since the measurement of an enzyme-catalyzed reaction is in actuality a measurement of a physiological effect of the enzyme interacting with its substrate, the relationships discussed in Chapters 7 and 8 for homotropic and heterotropic cooperativity are also valid when applied to enzyme kinetics. Since we measure reaction rates in place of equilibrium binding when dealing with enzymes, the spare receptor concept, the partial agonist concept, and the receptor desensitization concept have no exact correlates in enzyme kinetics. It is quite possible, however, for an enzyme to exist in more than one affinity state. This is similar to receptor desensitization.

Heterotropic and homotropic cooperativity as applied to enzyme kinetics will be discussed in Chapter 11. Enzymes can be inhibited or activated in the same ways in which receptors can be inhibited or activated. Therefore, most of the relationships discussed in Chapter 8 are also valid when applied to enzyme kinetics. There may be, under certain circumstances, some correlation between partial agonists functioning as partial competitive inhibitors of full agonists and substrates with high K_M values functioning as partial competitive inhibitors of substrates with low K_M values. However, the two situations are not precisely analogous.

Chapter 11

COMPLEX ENZYME-SUBSTRATE INTERACTIONS

I. INTRODUCTION

Complex enzyme kinetics can arise when an enzyme has more than one substrate, or more than one product, or both, for the reaction it catalyzes. These situations represent the rule rather than the exception. Greater than 60% of the enzymes, which have been characterized, fall into the category of having more than one substrate, or more than one product, or both. Generally, in an experimental setting, enzyme reactions that involve more than one substrate are set up with all but one of the substrates at saturating concentration. This often simplifies the system such that it behaves like a single substrate reaction. This does not always work, and sometimes valuable information can be overlooked by such simplification. Therefore, it is important to be able to recognize the effects of multiple substrates and/or multiple products on the kinetic behavior of enzymes.

Sticking with the nomenclature system established in Chapter 7 for multiple agonists, substrates will be designated as S_1, S_2, etc., with the subscript indicating the order of addition to the enzyme when the addition order is important. The products will be designated P_1, P_2, etc., in order of their release from the enzyme when the release order is important. Subscripts of the constants will conform to the subscripts of the substrates. Multiple products in sequential mechanisms will not alter the rate equations since we are dealing with measurements of the initial rate, where all products are at zero concentration. Mechanisms with more than two substrates will yield rate equations with correspondingly more terms to accommodate each of the substrates involved. The reader is referred to Cleland[13] for a full treatment of multiple substrate and multiple product reaction kinetics.

II. SEQUENTIAL, RANDOM ADDITION BI-UNI MECHANISM

The first mechanism to be considered here is for a reaction with two substrates and one product with random addition of the substrates.

$$\begin{array}{ccc} E + S_1 & \rightleftharpoons & ES_1 \\ + & & + \\ S_2 & & S_2 \\ \uparrow\downarrow & & \uparrow\downarrow \\ ES_2 + S_1 & \rightleftharpoons & ES_1S_2 \rightarrow E + P \end{array} \qquad \text{(Mechanism 20)}$$

As in Chapter 10, all possible transition forms of the ES_1S_2 and EP complexes have been combined into one term under the rapid equilibrium assumption. Mechanism 20 is a sequential, random addition bi-uni mechanism. The rate equation that describes it takes the same form as Equation 135 for noncompetitive activation of a receptor described in Chapter 8. In fact, often it is very difficult to distinguish between a second substrate and an enzyme activator or an enzyme cofactor.

The equation that describes Mechanism 20 is

$$\frac{V}{V_{max}} = \frac{1}{\frac{K_{M_2}}{[S_2}\left(1 + \frac{K_{D_1}}{[S_1]}\right) + \left(1 + \frac{K_{M_1}}{[S_1]}\right)} \tag{166}$$

K_{D_1} describes the interaction of E and S_1 to form ES_1. K_{M_1} describes the reaction ES_2 plus S_1 going to ES_1S_2 going to E plus P. K_{M_2} describes the reaction ES_1 plus S_2 going to ES_1S_2 going to E plus P. The K_M values from Mechanism 20, therefore, are equivalent to the K_M value for conventional Michaelis-Menten kinetics. Equation 166 can be rearranged to give the form of the equation more commonly seen in the enzyme kinetics literature:

$$V = \frac{V_{max}\,[S_1]\,[S_2]}{K_{D_1}K_{M_2} + K_{M_1}\,[S_2] + K_{M_2}\,[S_1] + [S_1]\,[S_2]} \tag{167}$$

Comparing Equation 167 with the Michaelis-Menten equation for a single substrate, single product mechanism (Equation 162), we see that there is an additional substrate term in the numerator and some extra terms in the denominator. V_{max} is the rate of the reaction when both substrates are saturating.

Equation 167 can be rearranged to give the double-reciprocal form of the equation

$$\frac{1}{V} = \frac{1}{[S_2]}\frac{K_{M_2}}{V_{max}}\left(1 + \frac{K_{D_1}}{[S_1]}\right) + \frac{1}{V_{max}}\left(1 + \frac{K_{M_1}}{[S_1]}\right) \tag{168}$$

Plotting 1/V vs. $1/[S_2]$ at various levels of $[S_1]$ yields a family of intersecting straight lines. When $K_{D_1} = K_{M_1}$ the lines intersect on the horizontal axis as shown in Figure 78. When $K_{D_1} > K_{M_1}$ the lines intersect above the horizontal axis as shown in Figure 79. When $K_{D_1} < K_{M_1}$ the lines intersect below the horizontal axis as shown in Figure 80.

When $[S_1]$ is at saturation the $K_{D_1}/[S_1]$ and $K_{M_1}/[S_1]$ terms become very small and drop out, simplifying Equation 166 to Equation 162. Since Mech-

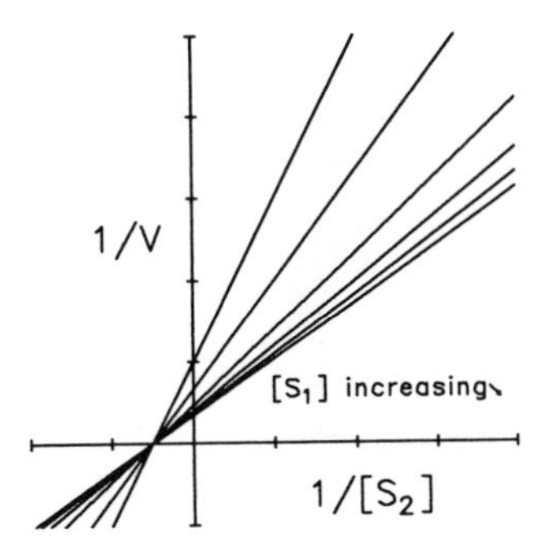

FIGURE 78

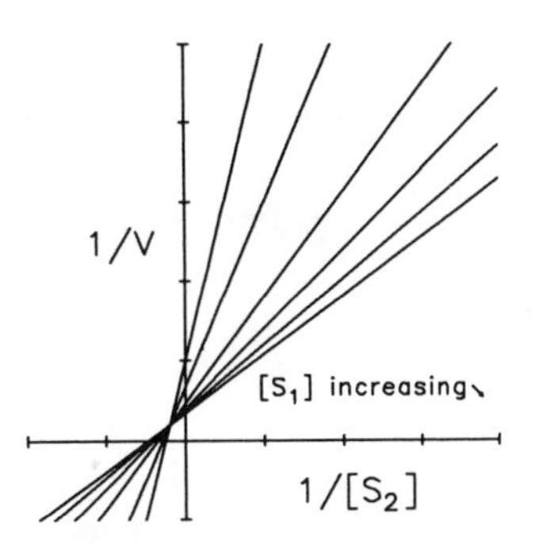

FIGURE 79

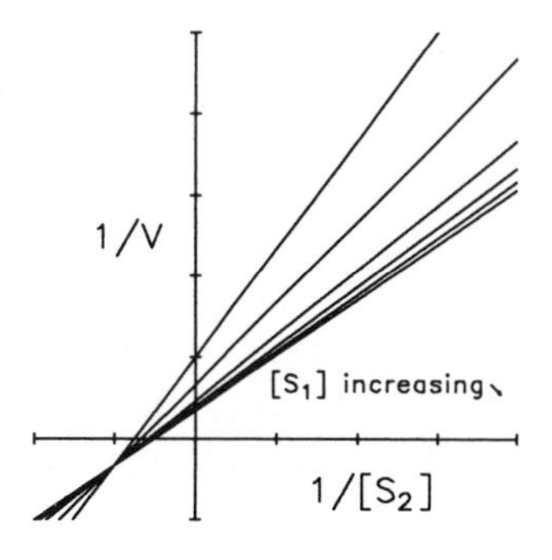

FIGURE 80

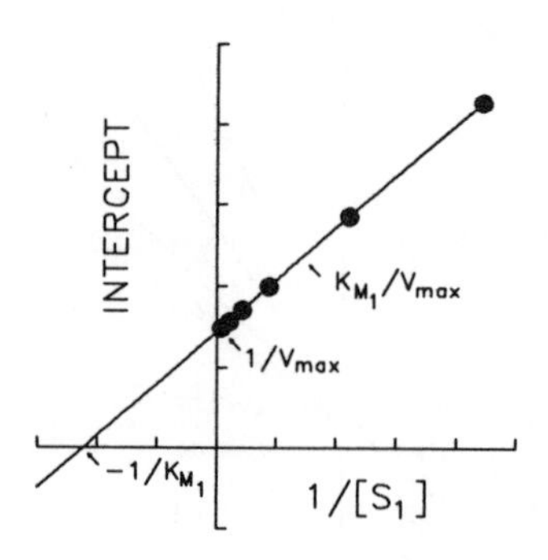

FIGURE 81

anism 20 is symmetrical for S_1 and S_2 it makes no difference which substrate is designated as 1 and which is designated as 2. Therefore, the K_M value for each substrate and the V_{max} can be determined unambiguously when the other substrate is saturating.

Alternatively, based on the relationship from Equation 168,

$$\text{intercept} = \frac{1}{V_{max}}\left(1 + \frac{K_{M_1}}{[S_1]}\right) = \frac{1}{[S_1]}\frac{K_{M_1}}{V_{max}} + \frac{1}{V_{max}} \tag{169}$$

it can be seen that a replot of the 1/V axis intercept vs. $1/[S_1]$, for data such as those in Figures 78 to 80, will give a straight line with slope K_{M_1}/V_{max}, ordinate intercept $1/V_{max}$, and abscissa intercept $-1/K_{M_1}$. This is shown in Figure 81. In addition, based on the relationship from Equation 168,

$$\text{slope} = \frac{K_{M_2}}{V_{max}}\left(1 + \frac{K_{D_1}}{[S_1]}\right) = \frac{1}{[S_1]}\frac{K_{D_1}K_{M_2}}{V_{max}} + \frac{K_{M_2}}{V_{max}} \tag{170}$$

it can be seen that a replot of the slopes vs. $1/[S_1]$ from data such as those for Figures 77 to 80 will give a straight line with slope $K_{D_1}K_{M_2}/V_{max}$, and ordinate intercept K_{M_2}/V_{max}. This is shown in Figure 82. Then, K_{M_2}/V_{max} (the intercept value from Figure 82) can be divided by $1/V_{max}$ (the intercept value from Figure 81) to obtain K_{M_2}. $K_{D_1}K_{M_2}/V_{max}$ (the slope value from Figure 82) can be divided by K_{M_2}/V_{max} (the intercept value from Figure 82) to obtain K_{D_1}.

III. SEQUENTIAL, ORDERED BI-UNI MECHANISM

Mechanism 20 can be altered to reflect a condition where the order of addition of substrates is important:

$$E + S_1 \rightleftharpoons ES_1 + S_2 \rightleftharpoons ES_1S_2 \rightarrow E + P \qquad \text{(Mechanism 21)}$$

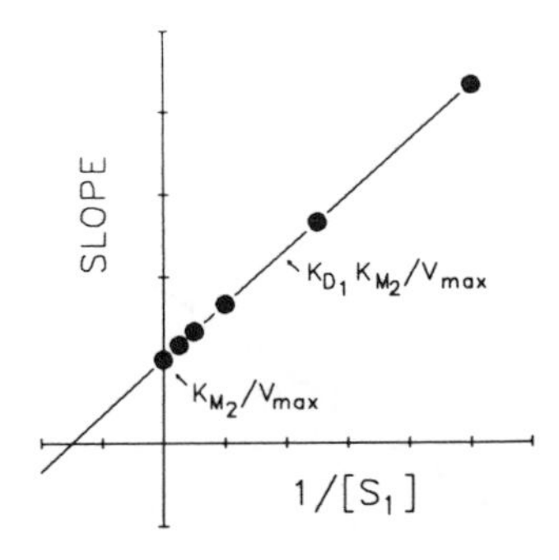

FIGURE 82

Mechanism 21 is a sequential, ordered bi-uni mechanism. In this mechanism the two substrates add to the enzyme in order, producing the ES_1S_2 complex. Then the product is produced and separates from the enzyme. The rate equation that describes Mechanism 21 takes the same form as Equation 167, which describes the sequential, random mechanism. In this mechanism, however, K_{M_1} is defined differently than K_{M_1} from Mechanism 20. Under special circumstances this difference in K_{M_1} can influence the graphic representation of the data. One instance where this can occur is in an ordered, sequential mechanism where S_1 functions as an activator for the enzyme rather than as a true substrate. The activator is not consumed by the reaction, as a true substrate would be. Therefore, it is not necessary for it to dissociate from the enzyme during a catalytic cycle. This behavior causes terms with K_{M_1} in them to drop out of the equations. This unbalances the equations with respect to S_1 and S_2. Such behavior is rare, however, and its absence is in no way an indication that the mechanism is random addition.

The most reliable means for distinguishing ordered, sequential mechanisms from random sequential mechanisms is by employing ligand binding measurements as described in Chapters 2 and 3. The enzyme and only one of the substrates, at several different concentrations, are combined and the amount of bound substrate is determined. The data thus generated will give the number of binding sites and the K_D value for the interaction of that substrate with the enzyme. If the mechanism is ordered sequentially, only S_1 will bind to the enzyme in the absence of the other substrate or substrates.

IV. PING-PONG MECHANISMS

To this point we have discussed only sequential mechanisms where all the substrates must add to the enzyme before any products can be released. Many enzymes have mechanisms where S_1 adds to the enzyme and P_1 leaves the enzyme before S_2 can add to produce P_2. The first reaction alters the enzyme in some way. For example, the enzyme may become oxidized by the

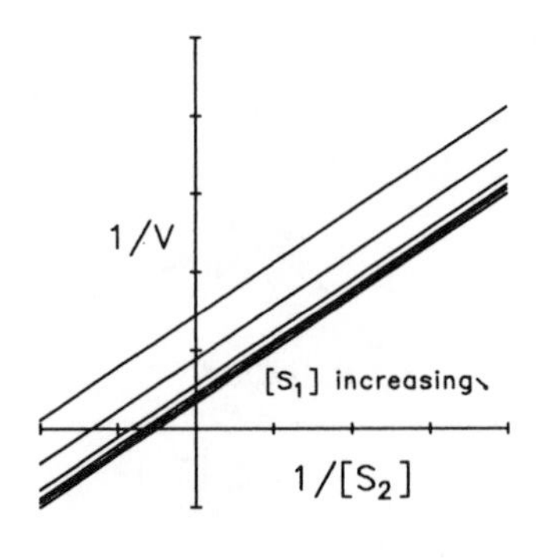

FIGURE 83

first reaction. Then S_2 adds to the altered enzyme and P_2 leaves. The second reaction reconverts the enzyme to its original form. This behavior is shown in Mechanism 22:

$$E + S_1 \rightleftharpoons ES_1 \rightharpoonup E^* + P_1$$

$$E^* + S_2 \rightleftharpoons E^*S_2 \rightharpoonup E + P_2 \qquad \text{(Mechanism 22)}$$

This is a ping-pong bi-bi mechanism. The rate equation that describes Mechanism 22 is

$$V = \frac{V_{max}\,[S_1]\,[S_2]}{K_{M_1}\,[S_2] + K_{M_2}\,[S_1] + [S_1]\,[S_2]} \tag{171}$$

Here K_{M_1} and K_{M_2} are defined in the same way as they were for Mechanism 20. We can see that the $K_{D_1}K_{M_2}$ term does not appear in the denominator of Equation 171 when compared with Equation 167. The double-reciprocal form of Equation 171 is

$$\frac{1}{V} = \frac{1}{[S_2]}\frac{K_{M_2}}{V_{max}} + \frac{1}{V_{max}}\left(1 + \frac{K_{M_1}}{[S_1]}\right) \tag{172}$$

Plotting 1/V vs. $1/[S_2]$ yields a family of parallel straight lines as shown in Figure 83. When $[S_1]$ is at saturation the $K_{M_1}/[S_1]$ term drops out of Equation 172. This simplifies Equation 172 to Equation 162. Again, since Equations 171 and 172 are symmetrical in regard to S_1 and S_2, and since Mechanism 22 could equally well have been written with the reaction involving E^* first, it doesn't matter which substrate is designated 1 and which is designated 2.

Comparison of Figures 77 to 80 with Figure 83 suggest that it should be easy to distinguish ping-pong mechanisms from sequential mechanisms. The problem with making mechanistic decisions from this form of analysis is that when $K_{D_1} >> K_{M_1}$ in a sequential mechanism, the intersection point may be so far to the left of zero on the horizontal axis that the lines will appear parallel.

V. COOPERATIVITY IN ENZYME KINETICS

There are several enzymes that have more than one catalytic site for the same substrate. Often these are enzymes with more than one subunit, where each subunit will have an active site. This circumstance is completely analogous to that discussed in Chapter 7 where multiple agonist binding sites exist on the same receptor complex. If these multiple binding sites behave independently of one another, the enzyme will exhibit simple Michaelis-Menten kinetics. If the multiple substrate sites interact with one another, the kinetics will exhibit cooperativity. Cooperativity in enzyme kinetics follows the same general rules as cooperativity in ligand binding and receptor kinetics. The equivalent equations and graphs are used to analyze for it.

There are many instances where enzymes may be inhibited, or activated, when their substrates also bind at separate inhibitor, or activator, binding sites. An example of this is the enzyme phosphofructokinase from the glycolytic pathway. This enzyme has two substrates, fructose-6-phosphate and ATP. ATP also functions as an inhibitor of this enzyme by binding to a binding site that is separate from the active ATP binding site. Substrate inhibition takes the same form as homotropic noncompetitive inhibition, which was described in Chapter 8. Substrate activation takes the same form as cooperativity, as described in Chapter 7.

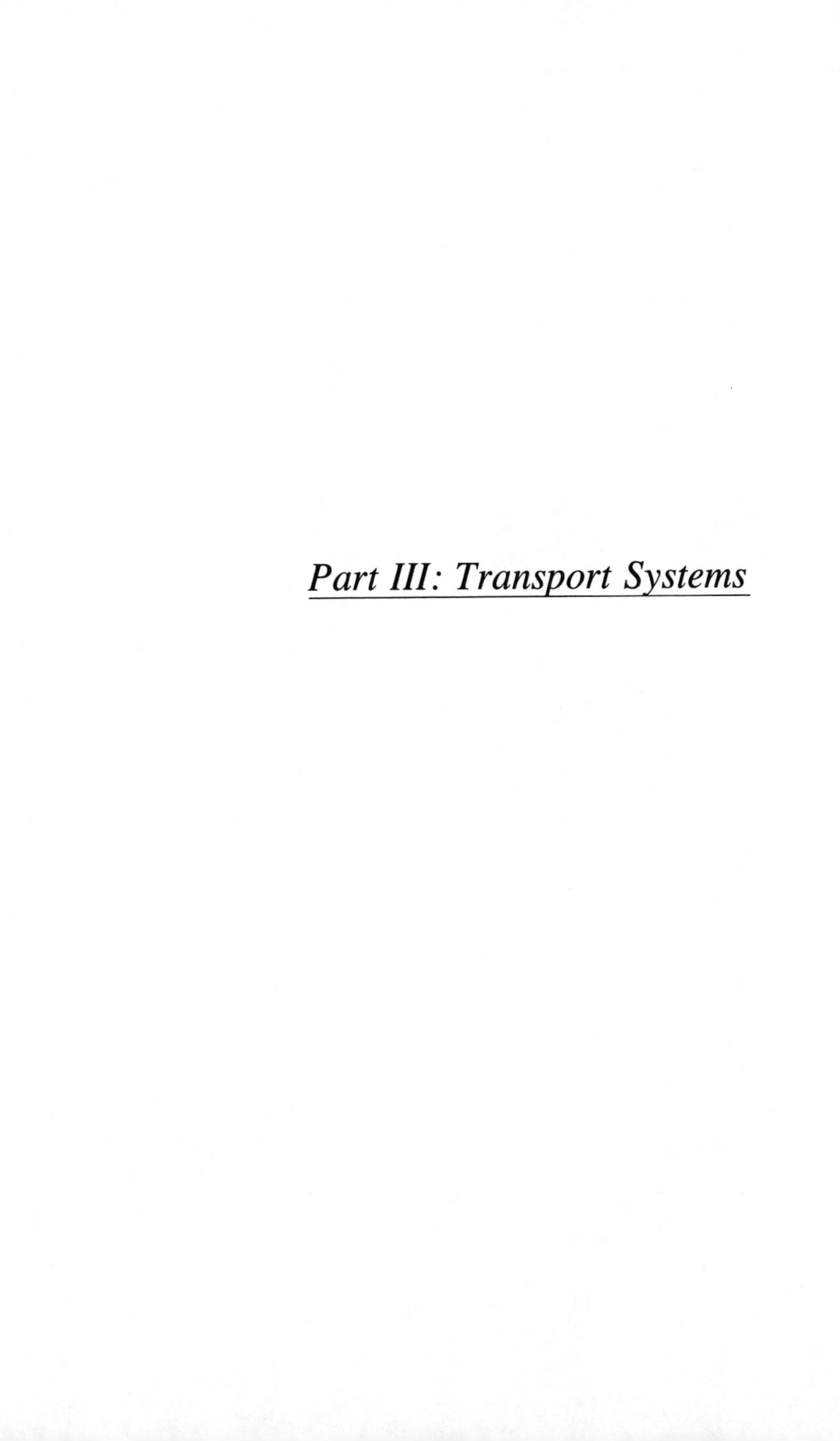

Part III: Transport Systems

Chapter 12

TRANSPORT KINETICS

I. INTRODUCTION

The final topic to be addressed in this book is that of transport kinetics. The term transport is used to refer to systems that facilitate the movement of solutes from one side of a membrane to the other. Nearly all the chemistry that occurs in living cells occurs at, in, or close to, a membrane. Cells are highly compartmentalized by membranous structures, and there is continuous need for chemical substances to be transported into or out of the various compartments. A simplistic, but useful, view of a cell is that it functions like a manufacturing facility. Raw materials must be continuously transported into the facility while waste and finished products must be continuously transported out. Within the facility, parts of the finished product are produced in specific locations and then transported to assembly sites, etc. Similar processes occur in cells. Nutrients must enter the cell from outside, while waste products and substances manufactured in the cell must be transported out of the cell. Many reactions carried out in the cell involve substances that are potentially dangerous to the cell, such as oxidative reactions. These must be sequestered within specialized compartments. Furthermore, there are processes such as electron transport and oxidative phosphorylation, for which membrane transport of materials is an essential part of the process.

Most of the substances that must cross membranes are not freely permeable to the membrane structure. Therefore, specific transport systems are present that not only facilitate the flow of materials from one side of the membrane to the other but also regulate the flow. Transport systems can be subclassified into two broad categories. Those requiring the input of energy are called active transport systems. Those that do not require the input of energy are called passive transport systems.

The overall tendency of matter is to go to a state of lowest possible free energy content. For many processes related to movement of solutes across a membrane this means going to the greatest state of disorder. If we have a two-compartment system with a membrane separating the compartments and we place water in one compartment and a sugar solution in the other compartment we have an orderly condition. To put this in less abstract terminology, suppose I have a jar full of marbles. The inside of the jar is one compartment and the outside of the jar is a second compartment. Having all the marbles in the jar is an orderly condition. If I overturn the jar the marbles will roll out and come to a more random distribution on the floor. Compared with the condition where all the marbles were in the jar, this is disorderly. This is analogous to having my clothing scattered all about my bedroom instead of neatly stored in my closet. It required very little effort, if any, on my part to distribute the marbles about the room on the floor. It takes considerable effort to collect all the marbles and get them back into the jar. Dumping the marbles out of the jar represents a passive form of transport. Getting them back into the jar requires the expenditure of energy and thus represents a form of active transport.

Returning to the example of a sugar solution on one side of a membrane, if there is a hole in the membrane the sugar will cross the membrane and redistribute itself on both sides. The redistribution will progress until there is no concentration difference across the membrane. This represents passive transport. Returning all the sugar to one side of the membrane requires the expenditure of energy. The movement of solutes from an area of low concentration to an area of high concentration is an energy-requiring, active transport process. The opposite is a passive transport process.

II. CHANNELS AND PORES

The transport of the sugar across the membrane was facilitated by making a hole in the membrane. One major subclass of passive transport systems does exactly this same thing, they make a hole in the membrane through which the solute can pass. This subclass includes channels and pore-forming substances. Ion channels and pore-forming antibiotics are good representative examples of such transport systems.

III. CARRIERS

Another group of transporters can be classified as carriers. Carriers are large molecules, generally proteins, which have a binding site for the substance to be transported that, when bound, is completely enclosed within the carrier. The external surface of the carrier is of a lipophilic nature, such that it can freely diffuse within the lipid core of a biological membrane. The carrier diffuses to the aqueous interface of the membrane where the solute binding site becomes exposed to the aqueous phase. This allows the solute molecule to enter the binding site. The carrier-solute complex then diffuses through the lipid core of the membrane to the other side, where the solute is released to the aqueous phase. When only one type of solute is being transported by the carrier, the direction of net solute flow will be from high concentration to low concentration. This is a form of passive transport.

IV. MEMBRANE FUSION

Membrane fusion systems involve enclosing the solute to be transported within a membrane vesicle. For transport into a cell this process is also known as phagocytosis, pinocytosis, or endocytosis. Portions of the cell membrane

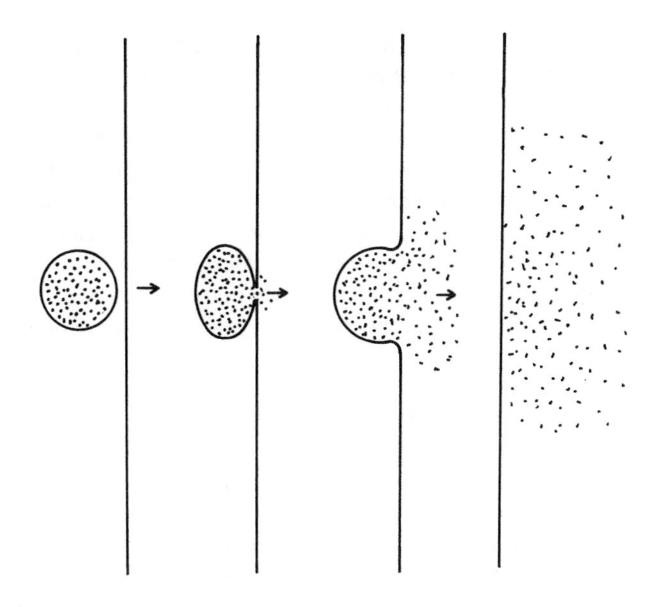

FIGURE 84

extend out to enclose the material to be transported into the cell. These portions of the membrane come together around the material, fuse together to become a continuous membrane and then separate from the cell membrane. In the process the solute has been brought into the cell. The opposite process (exocytosis) is used to move materials out of the cell. Membrane fusion-mediated transport is diagrammed in Figure 84.

V. ACTIVE TRANSPORT

A. ION PUMPS

Active transport systems are not as easily described. The best known example of an active transport system is the Na,K-dependent ATPase. This is a membrane-spanning enzyme that moves Na^+ from an area of low concentration inside the cell to an area of high concentration outside the cell. Simultaneously, K^+ is moved from an area of low concentration outside the cell to an area of high concentration inside the cell. The energy necessary to accomplish the transport of these ions is supplied by the hydrolysis of ATP. The ATPase enzyme, which accomplishes this active ion transport, has separate binding sites for Na^+ and K^+ and a catalytic site for ATP hydrolysis. Both Na^+ and K^+ are stimulators of the ATPase enzyme and both must be bound to the enzyme for the ATP hydrolysis to occur. The Na^+ binding sites are located on the inside surface of the membrane where they bind Na^+ with high affinity. The K^+ binding sites are located on the outside surface of the membrane where they bind K^+ with high affinity. Occupation of these sites stimulates the enzyme to hydrolyze ATP. The energy released by this hydrolysis is employed to cause the enzyme molecule to undergo a conformational change. The nature of the conformational change is such that the Na^+ binding sites are brought to the outside surface of the membrane while the K^+ binding sites are brought to the inside surface of the membrane. In addition, the conformational change causes alterations of the ion binding sites such that their affinities for their respective ions become markedly reduced. These affinity decreases allow the ions to dissociate from the binding sites even though they are now in areas of high concentration. Once the ion binding sites are unoccupied the enzyme reverts to its original conformation and the cycle can repeat.

B. SYMPORT AND ANTIPORT SYSTEMS

Many active transport systems operate on an antiport or symport basis. Antiport transporters move a solute molecule from an area of low concentration across the membrane to an area of high concentration. This is accomplished by also transporting a different solute molecule across the membrane in the opposite direction from an area of high concentration to one of low concentration. The energy for moving the substance from low to high concentration is supplied by the substance that is moving from high to low concentration.

Symporters move a molecule of solute across a membrane from an area of high concentration to one of low concentration while simultaneously moving a second solute across the membrane in the same direction from an area of low concentration to one of high concentration. Na^+ is often employed by such systems as the ion that moves downhill in concentration to provide the driving energy for the transport of the second solute.

VI. MEASUREMENT OF TRANSPORT

Transport of substances can be measured in several ways. Ions moving across membranes represent one form of electrical current. Ionic currents are associated with changes in the electrical potential (or voltage) across the membrane. Transmembrane electrical currents and voltages due to the operation of single ion channels can be measured. Such measurements represent the ultimate in our technological abilities to measure membrane transport. Electrical measurements are confined, however, to transport systems involving ions.

Alternatively, uptake and release of various substances can be measured with tissue slices, cells in culture, osmotically sensitive membrane vesicle preparations, and artificial membranes containing reconstituted transport systems. This last group of measurements is the true subject of the remainder of this chapter.

Other than electrical measurements, the most commonly employed techniques for measuring membrane transport involve monitoring the movement of a radiolabeled solute across a membrane. One adds the radiolabeled solute to one side of the membrane and observes the appearance of that solute on the other side of the membrane as time elapses. Often, such measurements will resemble radioligand binding measurements. If one measures influx of a solute into a membrane-enclosed space, such as a cell, one applies the radiolabeled solute on the outside of the membrane and at various time intervals collects the membranes, washes away the external radiolabel, and determines the amount of radiolabel that has crossed the membrane. Efflux measurements are accomplished in a similar fashion. First, the preparation is incubated with the radiolabeled solute for a sufficiently long period of time for it to equilibrate across the membrane. Then the membranes are washed in some way to remove the external radioligand. This is followed by measurements of the external solution for appearance of the radiolabel or collection and measurement of the membranes for loss of the radiolabel. When methods are available for directly detecting and measuring the solute being transported, similar techniques can be employed without the need for radiolabeling. This allows, for example, for the measurement of the release of solute endogenous to the preparation.

VII. ANALYSIS OF TRANSPORT DATA

With a few possible exceptions membrane transport involves, at some point, the interaction of a ligand with a binding site. This initiates the transport. Such systems, therefore, can be analyzed as receptor-effector systems. The classical case is a ligand-gated ion channel such as the nicotinic acetyl choline receptor-gated cation channel at the mammalian neuromuscular junction, which was described in Chapter 1. One can measure such receptor-agonist events through their effects on ion flow across a membrane.

Carriers and symport and antiport systems are somewhat further removed from classical receptor-effector systems, but they can still be analyzed as such. In these systems the solute (or ligand) must interact with a binding site on the transporter molecule to form a complex with it. This complex formation is the requisite first step in the transport of the molecule across the membrane. These systems are like enzymes, in some respects, and they are also like receptors in what they do.

All transport data can be analyzed as if they were enzyme rate data. Solute on the source side of the membrane represents the substrate and solute on the destination side represents the product. Transport K_M and V_{max} values can be found in the literature.

An alternative means of analyzing transport, which is mediated by ligand binding to a true receptor, is to measure equilibrated uptake or release of solute as a function of the concentration of the receptor ligand. Such an approach requires that several assumptions hold true in order that meaningful results can be obtained. The first assumption is that there must be a large number of individual compartments into which, or from which, the solute is transported. The second assumption is that there is only a very limited number of transport sites associated with each of the individual compartments. This assumption is best met in artificial vesicles containing reconstituted receptor-mediated transporters such that there is one, or only a few, transporter(s) per vesicle. Osmotically sensitive membrane vesicle preparations are the next best in meeting such an assumption since they can be derived from small cross-sectional areas of naturally occurring membranes. The third assumption is that the rate of flux of solute across the membrane is fast compared with the rate of dissociation of the receptor ligand from its binding site. This insures that the solute will equilibrate quickly compared with reequilibration of the receptor ligand. When these assumptions are met one has a system where the amount of solute that is released from, or sequestered by, the multiple small compartments is proportional to the fractional receptor occupancy. Data obtained for such systems can be analyzed in the same way as any other receptor-effector system data.

References

REFERENCES

1. **Langmuir, I.,** The absorption of gases on plane surfaces of glass, mica and platinum, *J. Am. Chem. Soc.*, 40, 1361, 1918.
2. **Michaelis, L. and Menten, M. L.,** Kinetics of invertase action, *Biochem. Z.*, 49, 333, 1913.
3. **Lineweaver, H. and Burk, D.,** The determination of enzyme dissociation constants, *J. Am. Chem. Soc.*, 56, 658, 1934.
4. **Rosenthal, H.,** A graphic method for the determination and presentation of binding parameters in a complex system, *Anal. Biochem.*, 20, 525, 1967.
5. **Scatchard, G.,** The attractions of proteins for small molecules and ions, *Ann. N.Y. Acad. Sci.*, 51, 660, 1949.
6. **Eadie, G. S.,** The inhibition of cholinesterase by physostigmine and prostigmine, *J. Biol. Chem.*, 146, 85, 1942.
7. **Hofstee, B. H. J.,** On the evaluation of the constants Vm and Km in enzyme reactions, *Science,* 116, 329, 1952.
8. **Woolf, B.,** cited in *Allgemeine Chemie der Enzyme,* by Haldane J. B. S. and Stern, K. G., Verlag von Steinkopff, T., Dresden and Leipzig, Germany, 1932, p 119.
9. **Henderson, L. J.,** The theory of neutrality regulation in the animal organism, *Am. J. Physiol.*, 21, 427, 1908.
10. **Hasselbalch, K. A.,** The calculation of the hydrogen number of the blood from the free and bound carbon dioxide of the same and the binding of oxygen by the blood as a function of the hydrogen number, *Biochem. Z.*, 78, 112, 1916.
11. **Hill, A. V.,** The possible effects of the aggregation of the molecules of haemoglobin on its dissociation curves, *J. Physiol. (London),* 40, 4, 1910.
12. **Schild, H. O.,** pA, a new scale for the measurement of drug antagonism, *Br. J. Pharmacol.*, 2, 189, 1947.
13. **Cleland, W. W.,** Steady state kinetics, in *The Enzymes,* Vol. 2, Boyer, P. D., Ed., Academic Press, New York, p. 1, 1970.

Appendices

Appendix 1

TABLE OF EQUATIONS AND THEIR USES

$$k = \frac{0.693}{t_{1/2}} \tag{4}$$

Equation 4 is useful for determining the value of the rate constant when the half-time is known or for determining the half-time when the rate constant is known. Rate constants and half-times are measured in dymamic binding experiments. The ratio of the dissociation rate constant to the association rate constant is an alternative way to determine the equilibrium dissociation constant. This relationship is only valid for first order processes or pseudo first order processes where all but one component is constant over time. Plotting the log of the concentration of the substance of interest vs. time will give a straight line if the process is first order.

$$\Delta G^\circ = -RT\ln K_D \tag{11}$$

Equation 11 is useful for determining the free energy difference between reactants and products at equilibrium when the equilibrium constant for the reaction is known. Alternatively, if the energy difference between reactants and products at equilibrium is known, Equation 11 can be used to determine the equilibrium constant. Calorimetric measurements can be used to measure ΔG° values associated with binding interactions.

$$B_L = \frac{B_{Lmax}\,[L]}{K_D + [L]} \tag{18}$$

The Langmuir binding isotherm[1] describes the characteristics of binding of molecules to other molecules or surfaces. We are concerned here with the binding of ligands to receptors. The useful parameters are the maximum binding capacity of the receptors (B_{max}) and their affinity for binding the ligand (K_D). Nonlinear fitting to Equation 18 is the best way of evaluating these parameters.

$$Y = \frac{[L]}{K_D + [L]} \tag{22}$$

An alternative means of expressing the Langmuir binding isotherm is in terms of fractional receptor occupancy (Y). This has the effect of confining the dependent variable to a scale from 0 to 1. This allows direct graphic comparisons (Y vs. log [L]) of the affinity of the interactions of different receptor-ligand systems. The ordinate scale is the same for each receptor-ligand interaction regardless of the magnitude of B_{max}. The only difference is the position of the curve, right or left, which is a measure of the affinity.

$$\frac{1}{B_L} = \frac{1}{[L]} \frac{K_D}{B_{Lmax}} + \frac{1}{B_{Lmax}} \tag{26}$$

The double-reciprocal or Lineweaver-Burk equation[3] is the most widely used and understood of the linear transformations of Equation 18. For this reason it is useful as a means of displaying data for graphic presentation. With the wide availability of computer programs for nonlinear fitting, Equation 26 is now of only limited value for evaluating binding parameters.

$$\frac{B_L}{[L]} = -\frac{B_L}{K_D} + \frac{B_{Lmax}}{K_D} \tag{29}$$

The Rosenthal-Scatchard equation[4,5] is particularly sensitive to the presence of heterogeneity of binding and for the presence of multiple binding sites for the same ligand on a single receptor. Either of these phenomena have the potential to produce nonlinearity. Multiple binding sites for the same ligand on a single receptor will produce convex curvature while binding site heterogeneity can produce convex, concave, or no curvature. The value of this transformation of Equation 18 is in graphic analysis for nonlinearity.

$$B_L = -\frac{B_L}{[L]} K_D + B_{Lmax} \tag{30}$$

The Eadie-Hofstee equation[6,7] is essentially the same equation as the Rosenthal-Scatchard equation. The major difference between the two equations is that the axes are reversed. Equation 30 is useful for the same analyses as the Rosenthal-Scatchard equation and it has the additional advantage of providing the values of the binding parameters directly rather than as reciprocals or in combinations.

$$\frac{[L]}{B_L} = \frac{[L]}{B_{Lmax}} + \frac{K_D}{B_{Lmax}} \tag{31}$$

The Woolf equation[8] is the least used of the linear transformations of the Langmuir binding isotherm. It is the best one for minimizing the effect of weighting errors. However, since nonlinear fitting to the Langmuir binding isotherm, as a means of evaluating binding parameters, is now widely available this feature of the equation is of diminished importance.

$$pL = pK_D + \log\left(\frac{[R]}{[RL]}\right) \tag{50}$$

Equation 50 is the ligand binding equivalent of the Henderson[9]-Hasselbalch[10] pH equation. This equation can be used like the Henderson-Hasselbalch equation, for example, to calculate the fractional receptor occupancy for any given ligand concentration when the K_D is known.

$$\frac{E_L}{E_{max}} = \frac{\alpha}{\frac{K_D}{[L]} + 1} \tag{61}$$

Equation 61 is a rearangement of Equation 23 that incorporates terms for relative intrinsic activity due to the phenomenon of partial agonism. When the maximum physiological effect for a receptor system is known, Equation 61 can be employed to evaluate the relative intrinsic activity (α) of any given agonist.

$$\log\left(\frac{Y}{1\text{-}Y}\right) = n \log [L] - \log K_D \tag{70}$$

The Hill equation[11] is useful for evaluating the presence of interacting binding sites on a receptor (cooperativity). The Hill equation also detects multiple noninteracting binding sites for the same ligand with different affinities on the same receptor. It is incapable of distinguishing between these two mechanisms. A Hill coefficient (n) greater than 1 is an indication of either cooperativity of binding or multiple noninteracting binding sites on the same receptor. The deficiency of the Hill equation is that the K_D value is a composite of multiple values and the Hill coefficient determined from experimental data, which is supposed to represent the number of different or interacting binding sites, is rarely a simple integer value.

$$\frac{B_L}{[L]^n} = -\frac{B_L}{K_D} + \frac{B_{Lmax}}{K_D} \tag{71}$$

Equation 71 is a form of the Scatchard equation that takes into consideration the possibility of multiple binding sites for the same ligand on the receptor. The n in this equation is the same as the Hill coefficient. When n is greater than 1 the Scatchard plot will exhibit pronounced convex curvature. This equation has the same deficiency as the Hill equation.

$$\log\left(\frac{Y}{1\text{-}Y}\right) = \log [L] - \log K_{D_2} - \log\left(1 + \frac{K_{D_1}}{[L]}\right) \tag{77}$$

Equation 77 is a modification of the Hill equation that takes into consideration the individual affinity constants for a receptor with two binding sites for the ligand where only the RLL complex is capable of producing a physiological effect. When these sites interact, or when they have different affinities, they will produce a downward curving Hill plot. When such behavior is noted this equation can be used to evaluate the relative contributions of the two binding sites to the overall binding and effect.

$$\log\left(\frac{Y}{1\text{-}Y}\right) = 2\log[L] - \log\left(K_{D_1}K_{D_2}\right) + \log\left(1 + \frac{K_{D_2}}{[L]}\right) \quad (82)$$

Equation 82 is a modified version of the Hill equation that takes into consideration the individual affinity constants for a receptor with two binding sites when both RL and RLL are capable of producing a physiological effect. This mechanism will produce Hill plots with an upward curvature. This equation, like Equation 77, is useful for evaluation of the relative contributions of the two binding sites to the overal binding and effect.

$$B_L = \frac{B_{Lmax}[L]}{K_D\left(1 + \frac{[I]}{K_I}\right) + [L]} \quad (94)$$

Equation 94 is a modified version of the Langmuir binding isotherm which takes into consideration the contribution of a competitive inhibitor. This equation, in combination with Equation 18, is useful for the evaluation of the inhibition constant employing nonlinear fitting methods.

$$\frac{1}{B_L} = \frac{1}{[L]}\frac{K_D}{B_{Lmax}}\left(1 + \frac{[I]}{K_I}\right) + \frac{1}{B_{Lmax}} \quad (95)$$

Equation 95 is a modified version of the double-reciprocal transformation of the Langmuir binding isotherm that takes into consideration the contribution of a competitive inhibitor. This equation is useful for the graphic analysis of competitive inhibition. If the inhibitor changes only the slope, and not the abscissal intercept, this indicates that the inhibition is competitive.

$$\log\left(\frac{[L']}{[L]} - 1\right) = \log[I] + pK_I \quad (103)$$

The Schild equation[12] is useful for calculating the concentration of agonist needed to restore the same level of effect when a known concentration of competitive inhibitor is added. Alternatively, it is useful for calculating the competitive inhibitor concentration when a known increase in agonist concentration is necessary to maintain the same level of effect.

$$E_L = \frac{E_{Lmax}\,[L]}{K_D\left(1 + \frac{[I]}{K_I}\right) + [L]\left(1 + \frac{[I]}{K_I}\right)} \tag{117}$$

Equation 117 is a modified version of the Langmuir binding isotherm which takes into consideration the contribution of a simple noncompetitive inhibitor. This equation is useful for the evaluation of the inhibition constant employing nonlinear fitting methods.

$$\frac{1}{E_L} = \frac{1}{[L]}\frac{K_D}{E_{Lmax}}\left(1 + \frac{[I]}{K_I}\right) + \frac{1}{E_{Lmax}}\left(1 + \frac{[I]}{K_I}\right) \tag{118}$$

Equation 118 is a modified version of the double-reciprocal transformation of the Langmuir binding isotherm that takes into consideration the contribution of a simple noncompetitive inhibitor. This equation is useful for the graphic analysis of simple noncompetitive inhibition. If the inhibitor changes both the slope and the abscissal intercept, this indicates that the inhibition is noncompetitive.

$$\frac{E_L}{E_{Lmax}} = \frac{[L]}{K_{D_1}\left(1 + \frac{[I]}{K_{I_1}}\right) + [L]\left(1 + \frac{[I]}{K_{I_2}}\right)} \tag{124}$$

Equation 124 is a modified version of the Langmuir binding isotherm which takes into consideration the contribution of a heterotropic-cooperative noncompetitive inhibitor. This equation is useful for the evaluation of the effect of agonist binding on the inhibition constant employing nonlinear fitting methods.

$$\frac{1}{E_L} = \frac{1}{[L]}\frac{K_{D_1}}{E_{Lmax}}\left(1 + \frac{[I]}{K_{I_1}}\right) + \frac{1}{E_{Lmax}}\left(1 + \frac{[I]}{K_{I_2}}\right) \tag{126}$$

Equation 126 is a modified version of the double-reciprocal transformation of the Langmuir binding isotherm that takes into consideration the contribution of a heterotropic-cooperative noncompetitive inhibitor. This equation is useful for the graphic analysis of heterotropic-cooperative noncompetitive inhibition. If the inhibitor changes both the slope and the abscissal intercept, this indicates that the inhibition is noncompetitive. If the intersection of the lines is above or below the ordinate, this indicates that the binding of substrate influences the binding of inhibitor and the binding of inhibitor influences the binding of substrate. When this is true the mechanism is heterotropic-cooperative noncompetitive inhibition.

$$Y = \frac{1}{\dfrac{K_{D_1}\left(1 + \dfrac{[I]}{K_{I_1}}\right)}{[L]\left(1 + \dfrac{[I]}{K_{I_2}}\right)} + 1} \qquad (128)$$

Equation 128 is a modified version of the Langmuir binding isotherm which takes into consideration the contribution of a heterotropic-cooperative non-competitive inhibitor that reduces the affinity of the receptor for the agonist without preventing the physiological effect. This is allosteric competitive inhibition. This equation is useful by virtue of the fact that it predicts that at very high inhibitor concentration there will be a decrease in the B_{Lmax}. Thus allosteric competitive inhibition can be distinguished from simple competitive inhibition by employing very high inhibitor concentrations.

$$Y = \frac{1}{\dfrac{K_{D_2}}{[L]}\left(1 + \dfrac{K_{S_1}}{[S]}\right) + \left(1 + \dfrac{K_{S_2}}{[S]}\right)} \qquad (135)$$

Equation 135 is a modified version of the Langmuir binding isotherm which takes into consideration the contribution of a simple noncompetitive stimulator. This equation is useful for the evaluation of the stimulation constant employing nonlinear fitting methods. An alternative way to think about a stimulator is that it can also represent,under some circumstances, a second agonist at the receptor where both agonists must bind before a physiological effect can be produced.

$$\frac{1}{E_L} = \frac{1}{[L]}\frac{K_{D_2}}{E_{Lmax}}\left(1 + \frac{K_{S_1}}{[S]}\right) + \frac{1}{E_{Lmax}}\left(1 + \frac{K_{S_2}}{[S]}\right) \qquad (136)$$

Equation 136 is a modified version of the double-reciprocal transformation of the Langmuir binding isotherm that takes into consideration the contribution of a simple noncompetitive stimulator. This equation is useful for the graphic analysis of simple noncompetitive stimulation. If the stimulator changes both the slope and the abscissal intercept, this indicates that the stimulation is noncompetitive.

$$\frac{E_L}{E_{Lmax}} = \frac{1}{\dfrac{K_D}{[L]} + \left(1 + \dfrac{[I]}{K_I}\right)} \qquad (144)$$

Equation 144 is a modified version of the Langmuir binding isotherm which takes into consideration the contribution of an uncompetitive inhibitor. This equation is useful for the evaluation of the inhibition constant employing nonlinear fitting methods.

$$\frac{1}{E_L} = \frac{1}{[L]}\frac{K_D}{E_{Lmax}} + \frac{1}{E_{Lmax}}\left(1 + \frac{[I]}{K_I}\right) \tag{145}$$

Equation 145 is a modified version of the double-reciprocal transformation of the Langmuir binding isotherm that takes into consideration the contribution of an uncompetitive inhibitor. This equation is useful for the graphic analysis of uncompetitive inhibition. If the inhibitor changes the abscissal intercept without affecting the slope, this indicates that the inhibition is uncompetitive. Uncompetitive inhibition can, under some circumstances, represent a special case of heterotropic-cooperative noncompetitive inhibition where binding of substrate increases the affinity of binding of the inhibitor from near zero to its final value.

$$V = \frac{V_{max}\,[S]}{K_M + [S]} \tag{162}$$

The Michaelis-Menten equation for enzyme kinetics[2] is equivalent in form to the Langmuir binding isotherm. A number of special conditions and assumptions are involved in the derivation of the Michaelis-Menten equation which are not necessary for the Langmuir binding isotherm. If these conditions and assumptions are not met the Michaelis-Menten equation is not valid for the analysis of enyme kinetic data. This equation is useful for the evaluation of the K_M and V_{max} parameters of enzyme reactions employing nonlinear fitting methods.

$$V = \frac{V_{max}\,[S_1]\,[S_2]}{K_{D_1}K_{M_2} + K_{M_1}\,[S_2] + K_{M_2}\,[S_1] + [S_1]\,[S_2]} \tag{167}$$

Equation 167 is a modified version of the Michaelis-Menten equation for enzyme reactions that require two substrates to add to the enzyme before any product formation can occur. Normally, one substrate is present at saturating concentration. When this is true the equation simplifies to Equation 162. This equation is useful for analyzing kinetic parameters when it is not feasible to hold one substrate at saturation.

$$V = \frac{V_{max}\,[S_1]\,[S_2]}{K_{M_1}\,[S_2] + K_{M_2}\,[S_1] + [S_1]\,[S_2]} \tag{171}$$

Equation 171 is a modified version of the Michaelis-Menten equation for enzyme reactions that exhibit a ping-pong mechanism. These enzymes have two substrates and two products. The first substrate combines with the enzyme and is converted to product. This reaction alters the enzyme such that it can then combine with the second substrate and convert it to product. The second reaction converts the enzyme back to its original form. Normally one of the substrates is present at saturating concentration. When this is true the equation simplifies to Equation 162. This equation is useful for analyzing kinetic parameters when it is not feasible to hold one substrate at saturation.

Appendix 2

SAMPLE CALCULATIONS

Part 1: Problems

Problem 1.
Two different enzymes catalyze the same chemical reaction (isozymes). Enzyme 1 has a K_M for the substrate of 5×10^{-7} *M* while enzyme 2 has a K_M for the substrate of 3×10^{-5} *M*. When the substrate concentration is 1×10^{-6} *M* what is the proportion of V_{max} for each enzyme?

Problem 2.
A compound which binds calcium ions has a K_D for calcium of 2.5×10^{-7} *M* at pH 7.4 and 25°C. What concentration of this compound would need to be added to reduce the free calcium ion concentration from 1×10^{-3} *M* to 5×10^{-5} *M*?

Problem 3.
You have two agonists which act at the same receptor site. Both agonists are full agonists (they produce the same maximal effect). Agonist 1 has a K_D of 7×10^{-7} *M* for the receptor while agonist 2 has a K_D of 2×10^{-8} *M*. What proportion of the maximal effect is produced when 7×10^{-7} *M* agonist 1 is mixed with with 2×10^{-8} *M* agonist 2?

Problem 4.
A similar problem to Problem 3 is suggested by the not uncommon practice of mixing agonists together to test whether the effects of the two agonists are synergistic. In order to determine whether the two agonists synergize one another we must first calculate their simple additive effects.

Using the same two agonists as in Problem 3, this time determine the proportion of maximal effect when each agonist is taken alone. For simplicity, choose concentrations of agonists which give the same proportion of maximal effect for each. Then combine the two agonists at one half of the original concentration for each. Is the effect of the combination of agonists equal to the effect of double the concentrations of each agonist alone at two times the concentration?

Problem 5.
You are characterizing the biological activity of a newly synthesized compound with potential value as a therapeutic agent. In a tissue bath preparation, which has been well characterized for analysis of muscarinic cholinergic activity, a 1×10^{-9} *M* concentration of this compound produces 25% of the known maximal effect for muscarinic agonists. If you assume that receptor

stimulation and physiological effect are proportional to receptor occupancy for this new compound, what is the K_D of the muscarinic receptors in the tissue preparation for this new agonist?

Problem 6.

A. You have an isolated smooth muscle preparation which contracts when an agonist drug is applied. This muscle is connected to a force transducer which allows you to measure the force of contraction. The maximum force of contraction that the preparation is capable of is 1.0 g. An agonist concentration of 1×10^{-8} *M* produces a contraction force of 0.75 g. What is the pL_2 value for this agonist?

B. You add a competitive antagonist at 5×10^{-7} *M*. This reduces the contraction force to 0.25 g. What is the pI_2 value for the antagonist?

C. What concentration of agonist will be necessary to achieve the original contraction force of 0.75 g in the presence of a 5×10^{-7} *M* antagonist?

Problem 7.

You have prepared a membrane suspension from a tissue homogenate which is enriched in the receptors that you desire to analyze. The membrane fragments can be collected by filtering through glass-fiber filters. You have a tritium-labeled receptor ligand (specific activity of labeling 7.99×10^{15} Bq/mmol — [^{3}H]-ligand), a supply of the same compound unlabeled (ligand), and a second unlabeled compound of a different chemical structure which binds to the same receptor site (displacer). You know from previous work that the affinities of the receptor for these compounds are in the 100-n*M* range. You have gone to the laboratory and obtained the binding data (radioactivity remaining on the filters) in Table 1.

Table 1
Receptor Binding Data

[[^{3}H]-Ligand] (n*M*)	[Ligand] (n*M*)	[Displacer] (n*M*)	DPM bound		
			Rep 1	Rep 2	Rep 3
1.0	0	0	270	285	259
2.5	0	0	673	681	650
5.0	0	0	1271	1368	1325
5.0	5.0	0	1343	1295	1256
5.0	20.0	0	1195	1147	1233
5.0	45.0	0	1138	1102	1077
5.0	95.0	0	882	888	861
5.0	245.0	0	638	598	616
1.0	0	100,000.0	25	27	22
2.5	0	100,000.0	65	62	60
5.0	0	100,000.0	121	124	123
5.0	5.0	100,000.0	123	125	122
5.0	20.0	100,000.0	120	123	124

Table 1 (continued)
Receptor Binding Data

[[^{3}H]-Ligand] (n*M*)	[Ligand] (n*M*)	[Displacer] (n*M*)	DPM bound Rep 1	DPM bound Rep 2	DPM bound Rep 3
5.0	45.0	100,000.0	126	122	122
5.0	95.0	100,000.0	123	124	121
5.0	245.0	100,000.0	125	124	122

Note: The volume of each assay sample is 1.0 ml.

A. Is there more than one detectable affinity for specific binding of agonist present in the membrane preparation?
B. What is (are) the affinity(s) of the receptors for the agonist?
C. How many receptor sites at each affinity are present?

Problem 8.
You have prepared a membrane suspension from a tissue homogenate which is enriched in the receptors that you desire to analyze. The membrane fragments can be collected by filtering through glass-fiber filters. You have a tritium-labeled receptor ligand (specific activity of labeling 7.99×10^{15} bq/mmol — [^{3}H]-ligand), a supply of the same compound unlabeled (ligand), and a second unlabeled compound of a different chemical structure, which inhibits the binding of the ligand (inhibitor). You know from previous work that the preparation shows a single affinity for the ligand in the 100-n*M* range. You have gone to the laboratory and obtained the binding data (radioactivity remaining on the filters) in Table 2.

Table 2
Receptor Binding Data

[[^{3}H]-Ligand] (n*M*)	[Ligand] (n*M*)	[Inhibitor] (n*M*)	DPM bound Rep 1	DPM bound Rep 2	DPM bound Rep 3
5.0	45.0	0	1087	1135	1084
5.0	57.5	0	1000	1030	1069
5.0	78.3	0	935	962	695
5.0	120.0	0	825	847	806
5.0	245.0	0	614	635	584
5.0	45.0	100.0	898	879	890
5.0	57.5	100.0	836	871	882
5.0	78.3	100.0	803	788	818
5.0	120.0	100.0	709	750	683
5.0	245.0	100.0	550	548	564
5.0	100,000.0	0	115	113	124
5.0	100,000.0	0	123	115	118

Note: The volume of each assay sample is 1.0 ml.

A. What are the K_D and B_{Lmax} of the preparation for the ligand?
B. What is the mechanism of inhibition?
C. What is the K_I value for the inhibitor?

Part 2. Solutions

Solution to Problem 1

To solve this problem we need the Michaelis-Menten equation for enzyme kinetics, Equation 162:

$$V = \frac{V_{max}\,[S]}{K_M + [S]} \tag{162}$$

Rearranging Equation 162 to an expression for the ratio of V to V_{max} will expedite the calculation.

$$\frac{V}{V_{max}} = \frac{[S]}{K_M + [S]} \tag{173}$$

Substituting in the values for substrate concentration and the K_M for enzyme 1, we can solve for V/V_{max}.

$$\frac{V}{V_{max}} = \frac{1 \times 10^{-6}}{5 \times 10^{-7} + 1 \times 10^{-6}} = 0.667$$

Repeating the calculation for enzyme 2 gives

$$\frac{V}{V_{max}} = \frac{1 \times 10^{-6}}{3 \times 10^{-5} + 1 \times 10^{-6}} = 0.032$$

From this calculation we can easily see that the enzyme with the smallest value of K_M will be operating at a much higher proportion of V_{max} when the substrate concentration is less than saturating.

Solution to Problem 2

This problem can be worked like a pH problem, using Equation 50

$$pL = pK_D + \log\left(\frac{[R]}{[RL]}\right) \tag{50}$$

What we want to know here is the amount of R necessary to complex all of the calcium ion except 5×10^{-5} *M*. pL is the minus log of the free calcium ion concentration, or the minus log of 5×10^{-5} *M*, which is 4.3; pK_D is 6.6. Substituting these values into Equation 50 gives

$$4.3 = 6.6 + \log\left(\frac{[\mathrm{R}]}{[\mathrm{RL}]}\right)$$

or

$$4.3 - 6.6 = -2.3 = \log\left(\frac{[\mathrm{R}]}{[\mathrm{RL}]}\right)$$

or

$$\frac{[\mathrm{R}]}{[\mathrm{RL}]} = 5 \times 10^{-3}\ M$$

We know that the total calcium concentration is $1 \times 10^{-3}\ M$ and that this is equal to the free calcium concentration, [L], plus the complexed calcium, [RL]. Therefore, [RL] is equal to total calcium minus free calcium. $[\mathrm{RL}] = 1 \times 10^{-3}\ M - 5 \times 10^{-5}\ M$, or $9.5 \times 10^{-4}\ M$. We can substitute this value for [RL] into Equation 176 to solve for [R].

$$\frac{[\mathrm{R}]}{9.5 \times 10^{-4}} = 5 \times 10^{-3}\ M$$

or $[\mathrm{R}] = 4.75 \times 10^{-6}\ M$. Then since the total [R] necessary is equal to [R] plus [RL], we add $4.752.3 \times 10^{-6}\ M + 9.5 \times 10^{-4}\ M$ to yield $9.55 \times 10^{-4}\ M$.

Problem 2 can also be worked with about the same amount of effort using the Langmuir binding isotherm as it appears in Equation 17.

$$[\mathrm{RL}] = \frac{[\mathrm{R_T}]\ [\mathrm{L}]}{\mathrm{K_D} + [\mathrm{L}]} \tag{17}$$

Here we want to solve for $[\mathrm{R_T}]$. From the above we know that [RL] is $9.5 \times 10^{-4}\ M$. Substituting for [RL], [L], and $\mathrm{K_D}$ into Equation 17 gives

$$9.5 \times 10^{-4} = \frac{[\mathrm{R_T}]\ 5 \times 10^{-5}}{2.5 \times 10^{-7} + 5 \times 10^{-5}}$$

rearranging and doing the arithmetic yields

$$[\mathrm{R_T}] = \frac{9.5 \times 10^{-4} \times 5.025 \times 10^{-5}}{5 \times 10^{-5}}$$

or

$$[\mathrm{R_T}] = 9.55 \times 10^{-4}\ M$$

Solution to Problem 3

To solve this problem we need Equation 113, the equation for the simultaneous effects of two agonists on the same receptor. This was discussed in Chapter 8.

$$\frac{E_{L_1L_2}}{E_{max}} = \frac{\alpha_1}{\frac{K_{D_1}}{[L_1]}\left[1 + \frac{[L_2]}{K_{D_2}}\right] + 1} + \frac{\alpha_2}{\frac{K_{D_2}}{[L_2]}\left[1 + \frac{[L_1]}{K_{D_1}}\right] + 1} \tag{113}$$

Since both agonists are full agonists α_1 and α_2 are both equal to 1. Also, since K_{D_1} is equal to $[L_1]$ and K_{D_2} is equal to $[L_2]$ all of the ratio terms in the equation become 1 and Equation 113 simplifies to

$$\frac{E_{L_1L_2}}{E_{max}} = \frac{1}{1(1 + 1) + 1} + \frac{1}{1(1 + 1) + 1}.$$

This further simplifies to

$$\frac{E_{L_1L_2}}{E_{max}} = 0.667.$$

A common mistake that is made is to simply add the fractional receptor occupancies together. Since both agonists were present at 50% fractional receptor occupancy, 50% plus 50% should equal 100%. From the above calculation it is easy to see how wrong this simplistic approach can be.

Solution to Problem 4

Again the alpha values are 1 for each agonist. Taking Equation 61, we can calculate the proportion of the maximal effect for each agonist by itself. Let us choose agonist concentrations of ten times the K_D value.

$$\frac{E_L}{E_{max}} = \frac{\alpha}{\frac{K_D}{[L]} + 1} \tag{61}$$

Substituting into Equation 61 for [L] equals ten times K_D gives

$$\frac{E_L}{E_{max}} = \frac{1}{\frac{1}{10} + 1} = 0.91$$

or, each agonist alone at ten times the K_D concentration gives 91% of the maximal effect.

Returning to Equation 113 and substituting in the values for [L] equal to five times K_D we have

$$\frac{E_{L_1L_2}}{E_{max}} = \frac{1}{\frac{1}{5}\left(1 + \frac{5}{1}\right) + 1} + \frac{1}{\frac{1}{5}\left(1 + \frac{5}{1}\right) + 1}$$

This simplifies to

$$\frac{E_{L_1L_2}}{E_{max}} = 0.91$$

Now let us see whether this relationship holds for any level of effect. This time choose [L] equals two times K_D. Equation 61 gives

$$\frac{E_L}{E_{max}} = \frac{1}{\frac{1}{2} + 1} = 0.667$$

Then, for the combination where the [L] concentrations are one times K_D we have

$$\frac{E_{L_1L_2}}{E_{max}} = \frac{1}{1(1 + 1) + 1} + \frac{1}{1(1 + 1) + 1} = 0.667$$

From this excercise we see that if there is no synergistic action the combination of agonists should give the same maximal effect as either agonist alone at twice the dose.

Solution to Problem 5

Because of our assumption of proportionality between receptor stimulation and physiological effect with receptor occupancy, we know that the fractional receptor occupancy (Y) produced by the new agonist is 25% or 0.25. Recalling the form of the Langmuir binding isotherm expressed in terms of fractional receptor occupancy we have

$$Y = \frac{[L]}{K_D + [L]} \tag{22}$$

Substituting for the values of [L] and Y gives

$$0.25 = \frac{1 \times 10^{-9}}{K_D + 1 \times 10^{-9}}$$

Rearranging and solving for K_D gives 3 x 10^{-9} *M*.

The only problem is that we have no way to assess whether or not the assumptions of proportionality were accurate. If they were not, the value obtained for K_D is inaccurate.

Solution to Problem 6

A. Recalling Equation 39, a rearrangement of the Langmuir binding isotherm in terms of effect ratio, we have

$$\frac{K_D}{[L]} = \frac{E_{Lmax}}{E_L} - 1 \tag{39}$$

E_L is 0.75 g and E_{Lmax} is 1.0 g. Substituting these values and the value for [L] (1×10^{-8} *M*) into Equation 38 gives

$$\frac{K_D}{1 \times 10^{-8} M} = \frac{1.0 \text{ gram}}{0.75 \text{ gram}} - 1$$

then solving for K_D yields $K_D = 3.33 \times 10^{-9}$ *M*. Taking the negative log gives $pK_D = 8.48$. Then from Equation 44, when receptor stimulation is proportional to receptor occupancy, and when effect is proportional to receptor stimulation, $pK_D = pL_2 = 8.48$.

B. We now need an equation for competitive inhibition in terms of effect ratio. The appropriate equation is

$$\frac{E_L}{E_{Lmax}} = \frac{1}{\frac{K_D}{[L]}\left(1 + \frac{I}{K_I}\right) + 1} \tag{173}$$

E_L is now 0.25 g and [L] is still 1×10^{-8} *M*, [I] is 5×10^{-7} *M*, K_D is 3.33×10^{-9} *M* (from part A), and E_{Lmax} is 1.0 g. Substituting all these values into the equation and solving for K_I yields

$$\frac{0.25 \text{ gram}}{1.0 \text{ gram}} = \frac{1}{\frac{3.33 \times 10^{-9} M}{1 \times 10^{-8} M}\left(1 + \frac{5 \times 10^{-7} M}{K_I}\right) + 1}$$

or

$$0.25 = \frac{1}{0.333\left(1 + \frac{5 \times 10^{-7} M}{K_I}\right) + 1}$$

Rearranging to solve for K_I gives

$$0.333\left(1 + \frac{5 \times 10^{-7}M}{K_I}\right) = \frac{1}{0.25} - 1 = 3$$

or

$$1 + \frac{5 \times 10^{-7}M}{K_I} = \frac{3}{0.333} = 9$$

or

$$0.333\left(1 + \frac{5 \times 10^{-7}M}{K_I}\right) = \frac{1}{0.25} - 1 = 3$$

and $K_I = 6.25 \times 10^{-8}\,M$, $pK_I = 7.2 = pI_2$.

C. Returning to Equation 173 we have

$$\frac{E_L}{E_{Lmax}} = \frac{1}{\frac{K_D}{[L]}\left(1 + \frac{I}{K_I}\right) + 1} \tag{173}$$

This time E_L is 0.75 g, E_{Lmax} is 1.0 g, [I] is $5 \times 10^{-7}\,M$, K_I is $6.25 \times 10^{-8}\,M$ (from part B) and K_D is $3.33 \times 10^{-9}\,M$ (from part A). Substituting these values into Equation 173 yields

$$\frac{0.75 \text{ gram}}{1.0 \text{ gram}} = \frac{1}{\frac{3.33 \times 10^{-9}M}{[L]}\left(1 + \frac{5 \times 10^{-7}M}{6.25 \times 10^{-8}M}\right) + 1}$$

or

$$0.75 = \frac{1}{\frac{3.33 \times 10^{-9}M}{[L]}(9) + 1}$$

and $[L] = 9 \times 10^{-8}\,M$.

Solution to Problem 7

Take the average DPM value of the three replicates for each treatment. For those treatments with 5 n*M* tritiated agonist and 100,000 n*M* displacer an average of all the samples is better. Then subtract the average value for the samples with 100,000 n*M* displacer (nonspecific bound, NSB) from the same ligand concentration with no displacer. This gives the specific bound.

Table 3

$[[^3H]$-Ligand] (n*M*)	[Ligand] (n*M*)	[Displacer] (n*M*)	$[[^3H]$-Ligand] average (DPM)	$[[^3H]$-Ligand] sp. bound (DPM)
1.0	0	0	271	246
2.5	0	0	668	606
5.0	0	0	1321	1198
5.0	5.0	0	1298	1175
5.0	20.0	0	1192	1069
5.0	45.0	0	1106	983
5.0	95.0	0	877	754
5.0	245.0	0	617	494
1.0	0	100,000.0	25	
2.5	0	100,000.0	62	
5.0	0	100,000.0	123	
5.0	5.0	100,000.0	123	
5.0	20.0	100,000.0	123	
5.0	45.0	100,000.0	123	
5.0	95.0	100,000.0	123	
5.0	245.0	100,000.0	123	

Next, convert the DPM for specific bound to n*M* $[^3H]$-ligand bound. To do this we need the specific activity of the labeled ligand (7.99×10^{15} Bq/*M*) and the conversion for DPM to bq (60 Bq/DPM). Dividing 7.99×10^{15} by 60 gives 1.33×10^{14} DPM/m*M*, and dividing this by the factor 1×10^6 n*M*/m*M* gives 1.33×10^8 DPM/n*M*. Then since the volume of each assay is 1 ml (0.001 l), and the ligand values are in concentration terms (n*M* or n*M*/l) we must multiply this factor times 0.001 l to obtain 1.33×10^5 DPM·l/n*M*. The values for specific bound in DPM are divided by this conversion factor to convert them to concentration of $[^3H]$-ligand bound in n*M*.

Table 4

$[[^3H]$-Ligand] (n*M*)	[Ligand] (n*M*)	$[[^3H]$-Ligand] sp. bound (DPM)	$[[^3H]$-Ligand] sp. bound (n*M*)
1.0	0	246	0.00185
2.5	0	606	0.00456
5.0	0	1198	0.00901
5.0	5.0	1175	0.00883
5.0	20.0	1069	0.00804
5.0	45.0	983	0.00739
5.0	95.0	754	0.00567
5.0	245.0	494	0.00371

The last column in Table 4 is then converted to total ligand bound in n*M*. To do this, first add the values in columns 1 and 2 together and divide by the value in column 1. This gives the labeled ligand dilution factor. Then multiply the corresponding value in column 4 times the ligand dilution factor to give the total ligand specific bound.

Table 5

[[^{3}H]-Ligand] (n*M*)	[Ligand] (n*M*)	[[^{3}H]-Ligand] sp. bound (n*M*)	Label dil. factor	Total [Ligand] sp. bound (n*M*)
1.0	0	0.00185	1	0.00185
2.5	0	0.00456	1	0.00456
5.0	0	0.00901	1	0.00901
5.0	5.0	0.00883	2	0.0177
5.0	20.0	0.00804	5	0.0402
5.0	45.0	0.00739	10	0.0739
5.0	95.0	0.00567	20	0.113
5.0	245.0	0.00371	50	0.186

Next, the concentration of unbound ligand must be calculated. This is necessary since the value for [L] in all of the binding equations is unbound (or free) [L] and not total [L]. Under normal experimental conditions the amount of ligand bound is a small fraction of the total ligand and unbound [L] will be nearly equal to total [L]. However, the correction is worth doing. This correction is done by subtracting the total amount bound (specific bound plus nonspecific bound) from the total ligand concentration. The most straightforward way to do this is to begin with the total ligand bound in DPM from Table 3 and convert these values to n*M* total ligand bound as we did for the values in Tables 4 and 5. Each of the DPM values for total bound from Table 3 is multiplied by the label dilution factor from Table 5 and then divided by 1.33×10^5 DPM·l/n*M* to convert them to total concentration bound. The values for total ligand bound are then subtracted from the total ligand in the assay to obtain the total unbound in n*M*.

Table 6

[[^{3}H]-Ligand] (n*M*)	[Ligand] (n*M*)	[Ligand] total (n*M*)	[Ligand] bound (n*M*)	[Ligand] unbound (n*M*)
1.0	0	1.0	0.00204	0.998
2.5	0	2.5	0.00502	2.495
5.0	0	5.0	0.00993	4.990
5.0	5.0	10.0	0.0195	9.980

Table 6 (continued)

$[[^3H]\text{-Ligand}]$ (nM)	[Ligand] (nM)	[Ligand] total (nM)	[Ligand] bound (nM)	[Ligand] unbound (nM)
5.0	20.0	25.0	0.0448	24.96
5.0	45.0	50.0	0.0832	49.92
5.0	95.0	100.0	0.132	99.87
5.0	245.0	250.0	0.232	249.8

The analysis we have accomplished so far is summarized in Figure 85. The top curve is a plot of the total ligand bound in nM vs. the total ligand concentration in nM. The bottom curve is a plot of the nonspecific bound vs. the total ligand concentration, and the middle curve is the specific bound ligand which was obtained by subtracting the bottom curve from the top curve.

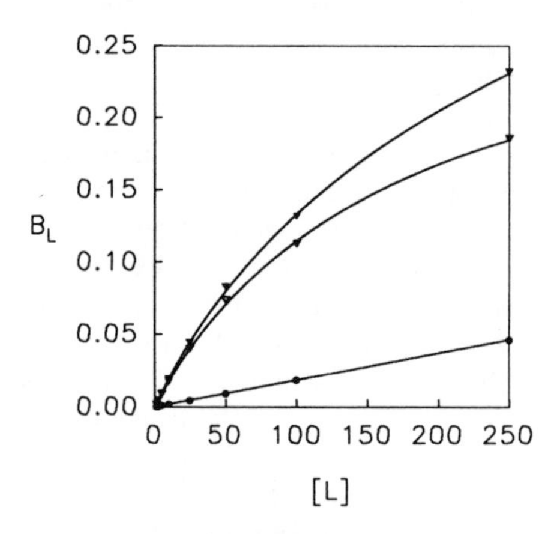

FIGURE 85

Next, the values for unbound ligand concentration ([L]) and total ligand specific bound (B_L) are fitted to the Langmuir binding isotherm for one site and two sites to determine best fit. This requires a nonlinear fitting computer program. The author's fitting program, which uses a simplex method, obtained the values of 0.31 nM for specific bound and 170 nM for the K_D when fitting to one site. When fitting to two sites the program determined that B_{L_1} was 0.17 nM, B_{L_2} was 0.27 nM, K_{D_1} was 100 nM, and K_{D_2} was 780 nM. Both fits had similar sum of squares errors. Since the two B_L values for the two-site fit are not widely different from one another, and the sum of squares errors for the two fits were similar, we must go to an Eadie-Hofstee plot of the data for the final evaluation of which model best fits the data. The Eadie-Hofstee transformation is best for this analysis for the reasons that this transformation is sensitive to the presence of multiple binding affinities and evaluation of the affinity and concentration of binding sites is straightforward. The Scatchard transformation is nearly equivalent to the Eadie-Hofstee. The double-reciprocal transformation is less sensitive to multiple affinities. The main value of the linear transformations is in plotting the data for visualization. This can also aid in the evaluation of the data. This plot shows the data points as solid circles, the short dashed line is the computer-generated one-site fit, and the two dotted lines are the computer-generated two-site fit. The B_L axis units are n*M*. Examination of the Eadie-Hofstee plot suggests that a single straight line with some variance due do experimental error is the best fit. Based on this and the facts that the sum of squares errors were similar for both fits, meaning that the data fit a one- or two-site model equally well and that the two B_L values for the two-site fit are fairly close together, leads to the conclusion that there is only one site.

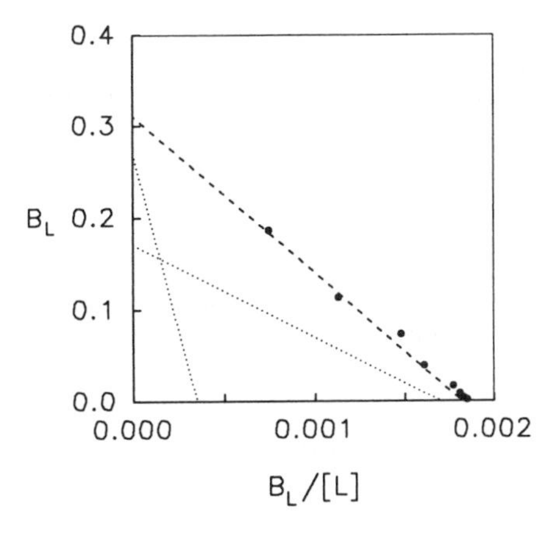

FIGURE 86

Solution to Problem 8

Take the average DPM value of the three replicates for each treatment. Then subtract the average value for the samples with 100,000 n*M* ligand (nonspecific bound, NSB) from the rest of the treatments. This gives the specific bound.

Table 7

$[[^3H]$-Ligand] (nM)	[Ligand] (nM)	[Inhibitor] (nM)	$[[^3H]$-Ligand] bound (DPM)	$[[^3H]$-Ligand] sp. bound (DPM)
5.0	45.0	0	1102	984
5.0	57.5	0	1033	915
5.0	78.3	0	954	836
5.0	120.0	0	826	708
5.0	245.0	0	611	493
5.0	45.0	100.0	889	771
5.0	57.5	100.0	863	745
5.0	78.3	100.0	803	685
5.0	120.0	100.0	714	596
5.0	245.0	100.0	554	436
5.0	100,000.0	100.0	118	
5.0	100,000.0	100.0	118	

Next, convert the DPM for specific bound to n*M* $[^3H]$-ligand bound. To do this we need the specific activity of the labeled ligand (7.99×10^{15} Bq/m*M*) and the conversion for DPM to bq (60 Bq/DPM). Dividing 7.99×10^{15} by 60 gives 1.33×10^{14} DPM/m*M* and dividing this by the factor 1×10^6 n*M*/m*M* gives 1.33×10^8 DPM/n*M*. Then, since the volume of each assay is 1 ml (0.001 l) and the ligand values are in concentration terms (n*M* or n*M*/l) we must multiply this factor times 0.001 l to obtain 1.33×10^5 DPM·l/n*M*. The values for specific bound in DPM are divided by this conversion factor to convert them to concentration of labeled $[^3H]$-ligand bound in n*M*.

Table 8

$[[^3H]$-Ligand] (nM)	[Ligand] (nM)	[Inhibitor] (nM)	$[[^3H]$-Ligand] sp. bound (nM)
5.0	45.0	0	0.00740
5.0	57.5	0	0.00688
5.0	78.3	0	0.00629
5.0	120.0	0	0.00532
5.0	245.0	0	0.00371
5.0	45.0	100.0	0.00580
5.0	57.5	100.0	0.00560
5.0	78.3	100.0	0.00515
5.0	120.0	100.0	0.00448
5.0	245.0	100.0	0.00328

The last column in Table 8 is then converted to total ligand bound in n*M*. To do this, first add the values in columns 1 and 2 together and divide by the value in column 1. This gives the labeled ligand dilution factor. Then multiply the corresponding value in column 4 times the ligand dilution factor to give the total ligand specific bound.

Table 9

[[^{3}H]-Ligand] (n*M*)	[Ligand] (n*M*)	[Inhibitor] (n*M*)	Label dil. factor	[Ligand] sp. bound (n*M*)
5.0	45.0	0	10.0	0.0740
5.0	57.5	0	12.5	0.0860
5.0	78.3	0	16.7	0.105
5.0	120.0	0	25.0	0.133
5.0	245.0	0	50.0	0.185
5.0	45.0	100.0	10.0	0.0580
5.0	57.5	100.0	12.5	0.0700
5.0	78.3	100.0	16.7	0.0860
5.0	120.0	100.0	25.0	0.112
5.0	245.0	100.0	50.0	0.164

Next, the concentration of unbound ligand must be calculated. This is necessary since the value for [L] in all of the binding equations is unbound (or free) [L] and not total [L]. Under normal experimental conditions the amount of ligand bound is a small fraction of the total ligand and unbound [L] will be nearly equal to total [L]. However, the correction is worth doing. This correction is done by subtracting the total amount bound (specific bound plus nonspecific bound) from the total ligand concentration. The most straightforward way to do this is to begin with the total ligand bound in DPM from Table 7 and convert these values to n*M* total ligand bound as we did for the values in Tables 8 and 9. Each of the DPM values for total ligand bound from Table 7 is multiplied by the label dilution factor from Table 9 and then divided by 1.33×10^5 DPM·l/n*M* to convert them to total concentration bound in n*M*. The values for total ligand bound are then subtracted from the total ligand in the assay to obtain the total unbound ligand in n*M*.

Table 10

[[^{3}H]-Ligand] (n*M*)	[Ligand] (n*M*)	[Inhibitor] (n*M*)	[Ligand] tot. bound (n*M*)	[Ligand] unbound (n*M*)
5.0	45.0	0	0.0838	49.9
5.0	57.5	0	0.0971	62.4
5.0	78.3	0	0.120	83.2
5.0	120.0	0	0.155	124.8

Table 10 (continued)

$[[^3H]$-Ligand] (n*M*)	[Ligand] (n*M*)	[Inhibitor] (n*M*)	[Ligand] tot. bound (n*M*)	[Ligand] unbound (n*M*)
5.0	245.0	0	0.230	249.8
5.0	45.0	100.0	0.0668	49.9
5.0	57.5	100.0	0.0811	62.4
5.0	78.3	100.0	0.101	83.2
5.0	120.0	100.0	0.134	124.9
5.0	245.0	100.0	0.208	249.8

Next, the values for unbound ligand concentration ([L]) and total ligand specific bound (B_L) for the treatments without inhibitor and, separately, for those with inhibitor are fitted to the Langmuir binding isotherm. This requires a nonlinear fitting computer program. Nonlinear fitting gives the values of 0.30 n*M* for B_{Lmax} and 150 n*M* for K_D for the uninhibited receptor, and 0.3 n*M* for B_{Lmax} and 210 n*M* for K_D for the inhibited receptor.

Since the B_{Lmax} values are the same for both the uninhibited and the inhibited cases, we can conclude that the inhibition mechanism is competitive. Recalling the equation for competitive inhibition, Equation 94

$$B_L = \frac{B_{Lmax}\,[L]}{K_D\left(1 + \frac{[I]}{K_I}\right) + [L]} \tag{94}$$

we see that in the inhibited case we have multiplied the K_D by the term $(1 + [I]/K_I)$. Remember that competitive inhibitors do not change the K_D value. We know that $K_D = 150$ n*M*, $[I] = 100$ n*M*, and from the data analysis and Equation 94 we know that

$$K_D\left(1 + \frac{[I]}{K_I}\right) = 210\ nM \tag{174}$$

Rearranging Equation 86 and substituting the values for K_D and [I] allows us to solve for K_I.

$$K_I = \frac{[I]}{\left(\frac{210\ nM}{K_D} - 1\right)}$$

or K_I equals 250 n*M*. The results of these calculations can be visualized by transforming the data to the double-reciprocal form and plotting.

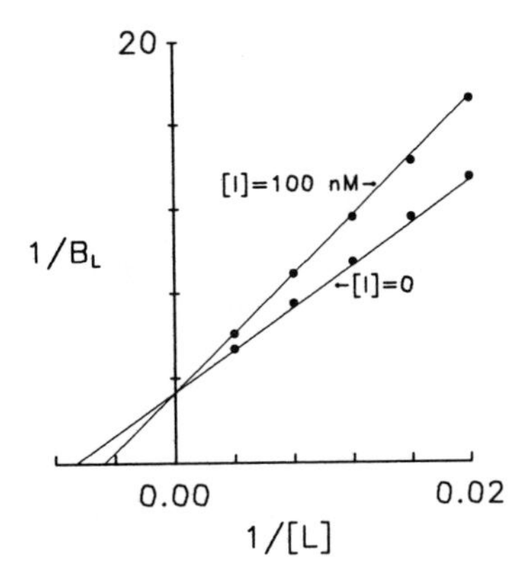

FIGURE 87

Index

INDEX

A

B

C

J

L

M

N

S